SOLVED PRACTICAL PROBLEMS IN
FLUID MECHANICS

SOLVED PRACTICAL PROBLEMS IN FLUID MECHANICS

Carl J. Schaschke

CRC Press
Taylor & Francis Group
Boca Raton London New York

CRC Press is an imprint of the
Taylor & Francis Group, an **informa** business

CRC Press
Taylor & Francis Group
6000 Broken Sound Parkway NW, Suite 300
Boca Raton, FL 33487-2742

First issued in paperback 2020

ISBN-13: 978-1-4822-4298-0 (hbk)
ISBN-13: 978-0-367-73789-4 (pbk)

Library of Congress Cataloging-in-Publication Data

Schaschke, Carl, author.
  Solved practical problems in fluid mechanics / Carl J. Schaschke.
     pages cm
  Includes bibliographical references and index.
  ISBN 978-1-4822-4298-0 (hardcover : acid-free paper) 1. Fluid mechanics--Problems, exercises, etc. I. Title.

TA357.3.S323 2015
620.1'06--dc23   2015025709

Visit the Taylor & Francis Web site at
http://www.taylorandfrancis.com

and the CRC Press Web site at
http://www.crcpress.com

Contents

Preface

The study of fluid mechanics forms an essential part of all engineering degree courses worldwide. While there have been many changes in education over the years, the teaching of mathematics and physical sciences remains critical to ensure that the next generation of engineers is fully equipped with all the necessary tools for professional practice. Providing undergraduates with solved problems has proved to be a successful and effective part of the learning process to demonstrate the complexities of the discipline.

This book is a comprehensive collection of problems with accompanying solutions in fluid mechanics which demonstrate the application of fluid flow principles in a range of commonly used engineering applications. It is a compilation of problems presented in a form that has a consistent nomenclature recognisable to all students of engineering. Aimed primarily at undergraduate students in the early stages of their academic formation, this book is also aimed at academic tutors as well as practitioners who may encounter challenging problems in fluid mechanics perhaps for the first time. Recognising that many students learn most effectively by the use of solved problems, the book will also be useful in the preparation of exams.

Each problem begins with a statement. The solution that follows does not present mathematical derivations but instead presents a solution from an easily recognisable starting point. Both problem and solution are therefore presented to enable the reader to follow each step in the analysis in a way that could be realistically achieved in a tutorial situation. The nomenclature is designed to be familiar to all engineers.

A number of the problems have been provided by academics who are directly involved in teaching fluid mechanics, and by industrialists. The problems selected are illustrative of key concepts and the significance of their solutions. They are tailored in such a way that the identity of a particular university or company and the nature of its business is avoided.

The problems include two-phase and multi-component flow, viscometry and the use of rheometers, non-Newtonian fluids, as well as new and novel applications of classical fluid flow principles. Each problem has been prepared using the SI system of units throughout, but it is recognised that non-SI units are still widely used in many industries. Reference is made to commonly encountered units and conversions presented, where appropriate.

In the preparation of this book, I am indebted to the many people who have assisted by providing solved problems. In particular, I would like to express my sincere appreciation to Dr. Isobel Fletcher (University of Strathclyde, Glasgow, Scotland), Andrew Bell (University of Strathclyde), Andrew McGuire (University of Cambridge, United Kingdom), Dr. Ian Wilson (University of Cambridge, United Kingdom), Dr. Andy Durrant

(University of the West of Scotland, Paisley). My thanks to the editorial staff of Taylor & Francis/CRC Press, Dr. Gangadeep Singh, Hayley Ruggieri, and Linda Leggio. Finally, this book could not have been produced without the support of my wife, Melodie, and my daughters, Emily and Rebecca.

The text has been carefully checked. Any errors, omissions, misprints, or obscurities are entirely my own. I would be grateful to receive suggestions for improvements.

Carl Schaschke
Abertay University
Dundee, Scotland

Author

Carl Schaschke, Ph.D., is a chemical engineer and is head of the School of Science, Engineering, and Technology at Abertay University (Dundee, Scotland), having previously served as head of the Department of Chemical and Process Engineering at the University of Strathclyde (Glasgow, Scotland) for 8 years. Prior to pursuing a Ph.D. in chemical engineering, he worked in the nuclear reprocessing industry at Sellafield (Cumbria, United Kingdom). He is a full professor and teaches fluid mechanics to undergraduates, and his research interests are in the thermophysical measurement of substances under extreme pressure. He serves as a UK representative of the European Federation of Chemical Engineering Working Party on High Pressure Technology. Dr. Schaschke has published several books including the *Dictionary of Chemical Engineering* (Oxford University Press, 2014). He is a fellow of the Institution of Chemical Engineers. He is married with two daughters, Emily and Rebecca.

Introduction

Fluid mechanics concerns the behaviour of fluids when subjected to changes in pressure; the effects of frictional resistance; the flow through pipes, ducts, and restrictions; and the production of power. The study of the behaviour of fluids forms an integral part of the education of an engineer for which a sound understanding of fluids is critical for the cost-effective design and efficient operation of machines and processes, and which also includes the development and testing of theories devised to explain various phenomena.

Generally well known for the large number of concepts required to solve even the simplest of problems, it is essential for the engineer to possess a sound grasp of the many concepts encountered in fluid mechanics to attempt to solve even the most seemingly straightforward of problems. A full and lucid grasp of the basics is therefore essential if such concepts are to be applied correctly and meaningfully. It is also worth remembering, however, that the solutions are only as valid as the mathematical models and the experimental data used to describe fluid flow phenomena. There can be no substitute for an all around understanding and appreciation of the underlying concepts and the ability to solve or check problems from first principles.

For many students or those new to the subject, there is often difficulty in identifying the necessary and relevant information to solve problems. Students may also be hesitant in applying theories covered in their studies, resulting from either an incomplete understanding of the principles or due to a lack of confidence as the result of unfamiliarity with the subject matter. While some concepts are straightforward, unexpected difficulties can be encountered when seemingly similar or related simple problems require the evaluation of a different but associated variable. Although the solution may involve the same starting point, the route through to the final answer may be quite different. Finding a clear path to solving a particular problem may therefore not always be straightforward. There is a propensity that the student may dwell unnecessarily on a mathematical quirk as the direct result of the application of an incorrect or inappropriate formula that is entirely due to the manner in which the problem had been incorrectly approached and which is irrelevant to the subject.

It is recognised that students develop and use a variety of study methods that are dependent on their own personal needs, circumstances, and available resources. For many, a quicker and deeper understanding of the concepts and principles is achieved when a problem is provided with an accompanying solution. The use of problems with solutions is an established and widely used approach to self-study. By providing a clear and logical approach from a distinct starting point through defined steps, together with the relevant mathematical formulae and manipulation, the student is able to gain an appreciation of both the depth and complexity involved in reaching a practical solution.

While applications require the straightforward application of the many fundamental principles of fluid mechanics that were founded in the 17th and 18th centuries by scientists such as Bernoulli, Newton, and Euler, many of today's fluid mechanics problems are complex, nonlinear, three-dimensional, and transient. High-speed and powerful computers are increasingly used to solve complex problems, particularly in computational fluid dynamics (CFD). Problems involving multi-phase flow can require involved procedures that are based on underlying concepts but require the use of empirical correlations based on experimental evidence. The combined flow of water and air along a horizontal pipe, for example, is complicated by the relative amounts of each phase, their relative velocities, and their different properties of density and viscosity, as well as interfacial surface tension between the two.

There has been long history of developments in the understanding of multi-phase processing. From the flow pattern maps and empirical correlations of Baker in the 1950s to today's highly sophisticated approaches of computational fluid dynamics requiring considerable computing power, the most valuable references for today's applications in texts and journals span the best part of half a century. The reference list provided at the end of this book is intended to enable the reader to delve more deeply and to review critically and compare models, form considered judgments, and distinguish between postulated models in terms of their merit.

This book is intended principally to support understanding in fluid mechanics. It is intended to be of assistance in solving related problems that may be encountered in a wide range of applications. Through the use of defined problems, the book is designed to enable the student to become familiar with, and to grasp firmly, important concepts and principles in fluid mechanics. Simple mathematical approaches have therefore been employed, although it is assumed that the reader has a prior knowledge of basic engineering concepts. Readers should be able to recognise similarities with their own problems and by following the provided solution, be able to reach their own solutions. The book, however, is not intended to be a complete and authoritative course or substitute to full texts on the subject. This book is therefore aimed at engineers who already have an understanding of fluid flow phenomena gained elsewhere. It requires the knowledge and application of fundamental engineering concepts such as dimensionless numbers and a fluency in basic mathematical skills such as differential calculus and associated application of boundary condition for solutions.

Each of the nine chapters has been prepared specifically to enable the reader to develop a sufficient knowledge and understanding of the fundamentals encountered in engineering, and to gain confidence in their application. The book is therefore intended to enable the reader to have an appreciation and understanding of fluids. Each problem has been selected and developed specifically to make as clear as possible, and without ambiguity or oversimplification, the important concepts used in this field of study. The book is therefore intended for the reader to draw on his or her own practical experience and to develop a critical and constructive approach to tackling problems.

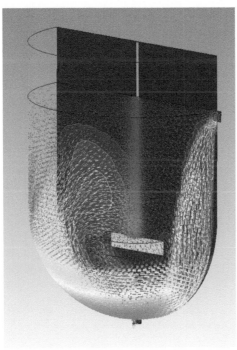

Nomenclature

Roman

A	Flow area	m^2		$\dot{m}$	Mass flow rate	kgs^{-1}
a	Coefficient	(—)		m_p	Mass of particle	kg
a	Lapse rate	$°\,km^{-1}$		N	Impeller diameter	m
B	Bingham number	(—)		N	Rotational speed	$rev\ s^{-1}$
b	Exponent	(—)		N_p	Power number	(—)
C	Concentration	$kmol\ m^{-3}$		N_s	Specific speed	$m^{3/4}s^{-3/2}$
C_d	Coefficient of discharge	(—)		P	Wetted perimeter	m
C_H	Head coefficient	(—)		P_o	Power number	(—)
C_Q	Capacity coefficient	(—)		p	Pressure	Nm^{-2}
c	Exponent	(—)		Δp	Pressure difference	Nm^{-2}
D	Impeller diameter	m		$\dot{Q}$	Volumetric flow rate	m^3s^{-1}
d	Diameter	m		R	Radius	m
d_c	Critical particle diameter	m		R	University gas constant	$kJkmol^{-1}K$
				r	Radius	m
d_e	Equivalent hydraulic diameter	m		r_H	Hydraulic radius	m
				Re	Reynolds number	(—)
d_p	Particle diameter	m		Re_B	Bingham Reynolds number	(—)
E_o	Eötvös number	(—)				
e	Voidage	(—)		S	Slip ratio	(—)
F	Force	N		S_n	Suction specific speed	(—)
f	Friction factor	(—)		s	Specific area	m^2m^{-3}
Fr	Froude number	(—)		T	Temperature	K
G	Mass flux	$kgm^{-2}s^{-1}$		T	Torque	Nm
g	Gravitational acceleration	ms^{-2}		t	Thickness	m
				t	Time	s
H	Depth	m		U	Fluid velocity	ms^{-1}
H_p	Head required for pumping	m		V	Volume	m^3
				W	Width	m
i	Gradient of slope	$m\ m^{-1}$		X	Multiplier	(—)
j	Superficial velocity	ms^{-1}		x	Horizontal distance	m
K	Coefficient	(—)		x	Mass quality	(—)
L	Length	m		y	Vertical distance	m
M_o	Morton number	(—)		Z	Compressibility	(—)
M_w	Molecular weight	(—)		z	Elevation	m

Greek

α	Gas void fraction	(—)
β	Coefficient	(—)
δ	Film thickness	m
ε	Surface roughness	m
η	Pump efficiency	(–)
θ	Angle of inclination	°
λ	Friction factor	(-)
λ_L	Liquid hold-up	(—)
μ	Viscosity	Nsm^{-2}
π	Pi	3.14 159
ρ	Density	kgm^{-3}
ρ_p	Particle density	kgm^{-3}
τ	Shear stress	Nm^{-2}
τ_w	Wall shear stress	Nm^{-2}
φ	Sphericity or shape factor	(—)
φ	Two-phase multiplier	(—)

1

Fluid Statics

Introduction

The term *general fluid* is applied to any substance that offers little or no resistance to change of shape to an applied force. Divided into three classes: liquids, vapours, and gases, it is liquids that offer the greatest resistance to compression and are not greatly affected by changes in temperature, whereas vapours and gases are easily compressed and are more susceptible to temperature changes. Solids may be made to behave as fluids in which they are dispersed as particles in liquids, vapours, or gases as in pneumatic conveying, which is a method of transporting solid particles such as grain and involves mixing the particles in a strong current of air. Toothpaste, pâte, and paints are fluids in which they retain their shape until an external force is applied causing them to flow, such as by way of a knife or brush.

The pressure of a fluid, whether moving or stationary, is the force it exerts over a given area or surface. Instrument gauges used to measure pressure express pressure as either absolute, which is measured above a total vacuum and absolute, or gauge, which is the pressure measured above atmospheric pressure, which itself is variable. The SI units are Newtons per square metre (Nm^{-2}) or pascals, abbreviated to Pa. Some gauges are calibrated using the Imperial units of pounds force per square inch (psi). Gauges used to measure a vacuum are expressed in torr. Absolute pressure refers to the pressure above total vacuum. Since the pascal is a small quantity, the term *bar* is used to represent 100,000 pascals (10^5 Pa). Standard atmospheric pressure is 1.013 bar or 14.7 psi (absolute).

In simple cases, the pressure within a static or stationary liquid may be considered to be incompressible, meaning that its density does not change appreciably with depth. The passenger liner RMS *Titanic*, for example, which sank in 1912 and which rests at a depth in the North Atlantic Ocean of around 4000 m, is subject to a pressure of around 400 bar. In contrast, the pressure of air above ground level is around 1.013 bar and decreases with elevation. The pressure at the top of the Matterhorn (4478 m) in the Alps is, for example, half the pressure than at ground level (Figure 1.1).

FIGURE 1.1
Matterhorn (4478 m) (Photo from C.J. Schaschke.)

Problem 1.1: Fluid Statics

Determine the pressure of the gas in the bulb shown in Figure 1.2 relative to the atmospheric pressure.

Solution

Pressure is the force applied to a fluid over a given area. Working from the gas bulb to the open end, a pressure balance based at the interface in each leg of the two "U" bends is

$$p_{gas} + \rho_m g h_1 = p_2 + \rho_m g h_2$$
$$p_2 + \rho_w g (h_3 - h_2) + \rho_m g h_3 = p_{atm} + \rho_m g h_4 \tag{1.1}$$

For the inclined leg

$$h_4 = L \sin 30 \tag{1.2}$$

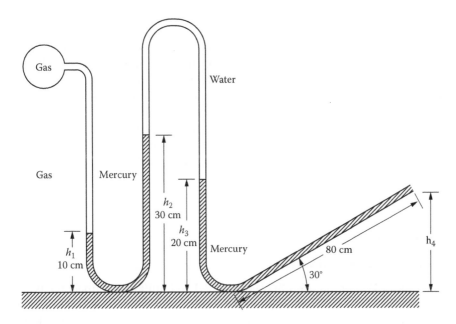

FIGURE 1.2
Variation of Static Pressure in a Tube

The difference in pressure between the bulb and the atmospheric pressure is therefore

$$p_{gas} - p_{atm} = g\left[\rho_m\left(h_2 - h_1 - h_3 + L\sin 30\right) - \rho_w\left(h_3 - h_2\right)\right]$$

$$= 9.81 \times \left[13,600 \times \left(0.3 - 0.1 - 0.2 + 0.8 \times 0.5\right) - 1000 \times \left(0.3 - 0.2\right)\right]$$

$$= 52,385\,\text{Pa} \tag{1.3}$$

This is about half an atmosphere above atmospheric pressure.

What Should We Look Out For?

Although the open part of the tube is inclined, the static pressure is measured vertically. In an inclined tube manometer, which is an instrument used for measuring the pressure head of a gas in which one leg of the manometer is attached to a sump containing a manometric fluid and the other is a straight tube usually made of glass and inclined at a known angle to the horizontal, the applied differential pressure between the sump and tube gives a vertical difference between the levels. The fluid displaced from the sump moves farther along the inclined tube than if the tube were vertical, thereby giving a multiplication effect improving the ability to measure small changes in pressure.

What Else Is Interesting?

The acceptable SI unit of pressure is the pascal, named after the French mathematician, physicist, and thinker on religion and philosophy, Blaise Pascal (1623–1662). Pascal, together with assistance from his brother-in-law, arranged an experiment on the Puy de Dome mountain in Auvergne, France, to demonstrate that the height of mercury in a barometer decreases with elevation. Pascal's pressure law states that the pressure applied to a fluid is transmitted equally in all directions and is also known as Pascal's principle. Pascal suffered ill health throughout most of his life and died at the early age of 39.

Problem 1.2: Falkirk Wheel

The Falkirk Wheel (Figure 1.3) was designed and built to reconnect the Forth & Clyde and Union Canals in central Scotland and is the only rotating boat lift in the world. It consists of two identical tanks to transport simultaneously watercraft between the 35 m difference in elevations. Explain why both tanks will always remain balanced even if one tank should contain two heavily laden barges while the other a single canoe.

Solution

The original idea and design of the Falkirk Wheel to act as a boat lift, dates back to the 19th century, although it was only considered as a solution for joining the two canals in 1994. It is an illustration of Archimedes' principle for floating bodies that states that when a body floats it displaces a weight, of liquid equal to its own weight. In fact, the principle was not stated by Archimedes but is connected to his discoveries in hydrostatics. When a body is partially or totally immersed in a liquid, there is an upthrust on the body equal to the weight of the liquid displaced by the body.

Both tanks will be perfectly balanced irrespective of the type, size, and number of floating vessels within each. This is because each tank displaces its own weight in water and therefore both sides will remain balanced.

What Should We Look Out For?

According to Archimedes' principle, a body will float in a liquid if the displaced liquid weighs less than the body itself. The hull of a ship made from steel will displace a volume or weight of water greater than the overall weight

FIGURE 1.3
Falkirk Wheel (Photo from Melodie Schaschke.)

of the ship itself. Ice floats freely on water since it displaces its own weight, which is equal to the weight of the water that is displaced. That is,

$$m_{ice}g = m_{water}g \tag{1.4}$$

Expressing mass in terms of volume and density, for which the density of ice is 911 kgm⁻³, the proportion below the surface, F, to the overall height, L, is therefore,

$$\frac{F}{L} = \frac{\rho_{ice}}{\rho_W} = \frac{911}{1000} = 0.911 \tag{1.5}$$

That is, over 90% of the ice lies below the surface.

What Else Is Interesting?

Archimedes of Syracuse (287–212 BC) was a Greek mathematician and philosopher. He is credited with the mechanics of levers, the Archimedean screw pump, and a method of successive approximations that allowed him to determine the value of π to a good approximation. It is said that King Hiero asked Archimedes to check whether a crown was made of pure gold throughout or contained a cheap alloy. While in a public bath and pondering on how to do this without damage to the crown, Archimedes is supposed to have suddenly thought of the possibility of immersing it in water and checking its density by way of displacement, and to have been so excited that he ran naked through the streets shouting Eureka! Eureka! I have found it! I have found it! He was killed by a soldier in the Roman siege of Syracuse.

Problem 1.3: Gauge Pressure

A water-filled bulb within a chamber is connected to the outside by two U-tube manometers as shown in Figure 1.4. Determine the gauge pressure in the bulb.

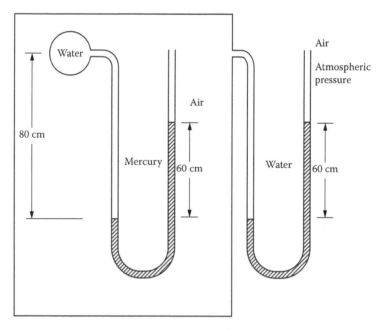

FIGURE 1.4
Connecting Manometers

Solution

The gauge pressure is the pressure measured relative to atmospheric pressure, which is variable. The pressure in the chamber is greater than that of the atmospheric pressure by virtue of difference in levels in the manometer. With the level of mercury being higher in the right-hand leg connecting the bulb, the gauge pressure in the bulb must be greater still. The pressure in the chamber is, therefore,

$$p_{chamber} = -\rho_w g h_1 = -1000 \times 9.81 \times 0.6 = -5886 \text{ Nm}^{-2} \tag{1.6}$$

while the pressure in the bulb is

$$p_{bulb} = p_{chamber} + \rho_{Hg} g h_1 - \rho_w g h_2 = -5888 +$$
$$13,600 \times 9.81 \times 0.6 - 1000 \times 9.81 \times 0.8 = 78,098 \text{ Nm}^{-2} \tag{1.7}$$

What Should We Look Out For?

A differential manometer is used to measure the difference in pressure between two points such as across a heat exchanger or filter system and is typically used to determine whether there is a blockage or blinding of the filter. A U-tube-type differential manometer consists of a U-tube containing a manometric fluid which is opaque with a higher density than the process fluid, inert, and has a low vapour pressure. The usual manometer calculations would include the difference in densities between the manometric fluid and the air. Where mercury is used as the manometric fluid, the density of air is less than 0.01% of the density of the mercury and can therefore usually be ignored. Reading the level of mercury in the glass U-tubes involves reading off the top of the meniscus to a graded scale on the glass.

What Else Is Interesting?

The term *negative pressure* is used to mean pressure below atmospheric pressure. Negative room pressure is a technique used to ensure that the flow of air is into a room and to prevent the release of air out. Applications include hospitals and medical centres in which negative pressure is used to prevent the release of potentially contaminated air from a room. Is it also practiced in the nuclear industry to prevent the release of potentially contaminated air. The negative pressure is created by a ventilation system that draws more air out of the room than is permitted to enter. For even a modest pressure difference across a door of 10 cm of water gauge, the force required to open a standard door with an area of 2 m² corresponds to nearly 2 kN.

Problem 1.4: Air Pressure with Altitude

Determine the variation of air pressure and temperature with altitude up to an altitude of 11 km if the ground temperature and pressure may be taken to be 15°C and 1013 mbar, respectively.

Solution

The atmospheric pressure of air, like a static liquid whose pressure varies with depth, varies with elevation. At ground level, standard atmospheric pressure is often conveniently taken as 101.325 kNm^{-3} and 288.15 K. Frequently referred to as 1013 millibar, there are daily pressure variations due to prevailing meteorological conditions with both highs and lows about this value.

The pressure exerted by the molecules in air is attributed to the gravitational attraction between the earth and the molecules. The pull of air by Earth's gravitational field was discovered in the 17th century. It was the Italian scientist Evangelista Torricelli (1608–1647) who appropriately concluded that "We live submerged at the bottom of an ocean of the element air" (West 2013). Torricelli went on to develop the mercury barometer in 1643, which is an instrument used to measure atmospheric pressure and its variations.

To determine the variation of air pressure with altitude, the gravitational acceleration is assumed to be constant at sea level. The variation of pressure with elevation for a static fluid is therefore

$$dp = -\rho g dh \qquad (1.8)$$

The negative sign indicates a reduction in pressure with rise in elevation. If the density of air can be assumed to behave as an ideal gas where

$$\rho = \frac{p}{RT} \qquad (1.9)$$

then the variation of pressure with elevation is

$$\frac{-dp}{dh} = \frac{pg}{RT} \qquad (1.10)$$

Known as the lapse rate, the temperature decreases linearly by approximately 6.5°C per vertical kilometre up to an altitude of 11 km. That is,

$$\frac{-dT}{dh} = a \qquad (1.11)$$

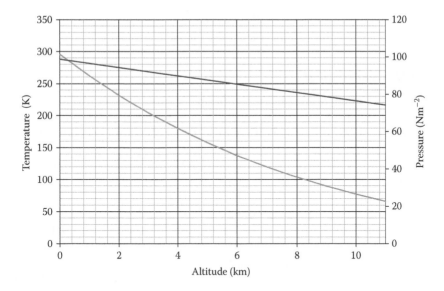

FIGURE 1.5
Variation of Temperature and Pressure with Altitude

The variation of pressure with temperature can therefore be determined from the integration:

$$\int_{p_S}^{p} \frac{dp}{p} = \frac{g}{Ra} \int_{T_S}^{T} \frac{dT}{T}$$ (1.12)

to give

$$p = p_s \left(\frac{T}{T_s} \right)^{\frac{g}{Ra}}$$ (1.13)

The variation of pressure with elevation is given in Figure 1.5.

Combining Equations 1.11 and 1.13 enables the altitude to be determined for a given pressure:

$$h = \frac{T_s}{a} \left(1 - \left(\frac{p}{p_s} \right)^{\frac{Ra}{g}} \right)$$ (1.14)

What Should We Look Out For?

The value for the universal gas constant, R, has the usual SI units of 8.314 kJ kmol^{-1} K^{-1}. In this case, the value is converted to 287.05 Jkg^{-1}K^{-1} since the mass of air is taken as a mixture of 21% oxygen and 79% nitrogen for which the average molecular mass is therefore taken to be 29 kmols/m^{-3}.

What Else Is Interesting?

Based on standard atmospheric conditions at ground level, the summit of Mount Everest (8848 m) corresponds to a pressure of 31.4 kNm^{-2} and a temperature of $-43°C$. As altitude increases, the air therefore becomes thinner and consequently contains less oxygen. This information is used to determine the effects of oxygen deprivation faced by mountaineers at high altitude. Not surprisingly, perhaps, many such mountaineers succumb to the effects of altitude sickness, frostbite, and hypothermia.

On aircraft, most flight instruments do not directly measure altitude but instead measure pressure and deduce altitude. For very obvious safety reasons of both the crew and passengers, commercial aircraft use pressurised cabins. The cabin pressure is carefully maintained at a safe level below atmospheric pressure, but above that which can cause noticeable effects of altitude symptoms. This corresponds to an equivalent altitude of around 3000 m even though the aircraft may be flying at a cruising attitude of 9000 m. Otherwise, thicker-walled aircraft to contain higher cabin pressure would require larger aircraft and additional fuel costs.

Problem 1.5: Pascal's Paradox

A vessel consists of a body with a cross-sectional area of 0.5 m^2 and a tube of area 0.005 m^2 as shown in Figure 1.6.

Solution

At the bottom of the vessel, the hydrostatic pressure is given by

$$p = \rho g h = 1000 \times 9.81 \times 1.0 = 9.81 \, \text{kNm}^{-2} \tag{1.15}$$

The force at the base of the vessel is therefore

$$F = pa = 9810 \times 0.5 = 4.905 \, \text{kN} \tag{1.16}$$

What Should We Look Out For?

The internal pressure is not due to the volume but due to the hydrostatic depth only. This presents a curiosity if one were to consider the downward

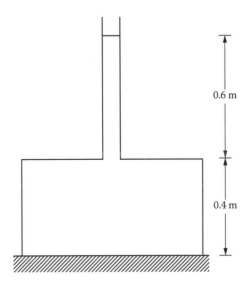

FIGURE 1.6
Pressure in a Vessel

force to the weight of liquid. Ignoring the weight of the vessel, the weight of the water is

$$F = \rho Vg = 1000 \times (0.5 \times 0.4 + 0.005 \times 0.6) \times 9.81 = 1991\,N \qquad (1.17)$$

What Else Is Interesting?

To illustrate the significance of hydrostatic depth, rather than shape of vessel, Pascal's vases are an interesting collection of vessels or "vases," which each filled to the same depth exhibit the same pressure at the bottom (Figure 1.7).

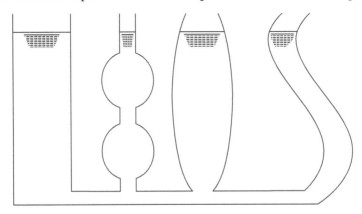

FIGURE 1.7
Pascal's Vases

The various "vases" are each linked to a communal reservoir. The equilibrium situation requires that each must have the same surface level.

The pressure with depth is the same irrespective of the shape of the vase.

Problem 1.6: Fish Ladder

Incorporated into the hydroelectric dam on the river Tummel in Perthshire, Scotland, is a fish ladder, which is designed to enable spawning salmon that head upriver to reach Loch Faskally to bypass the dam. The fish ladder consists of a series of pools or steps each connected by an underwater pipe through which the fish swim (Figure 1.8) and rise up 16 m from the river Tummel to reach the Loch above. If the fish are capable of swimming against a flow through the connecting pipes where the mean velocity is 3 ms^{-1}, determine the number of steps in the ladder.

Solution

Fish ladders, which are also known as fishways or fish passes, are ways in which fish are able to overcome obstacles placed across waterways such as dams and locks. They usually permit the fish to swim or leap their way around the obstacle in a series of low steps that collectively gives the name *fish ladder*. The type of fish ladder on the Tummel was a solution to a problem

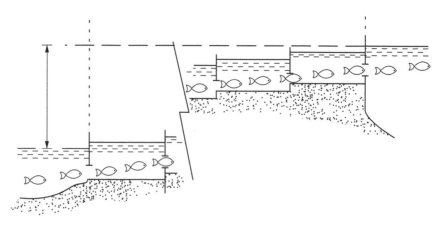

FIGURE 1.8
Fish Ladder

during the damming of the river at the Scottish Highland town of Pitlochry in the early 1950s in which migrating salmon and trout were able to bypass the obstacle of the dam and hydroelectric power station. It consists of a connected series of pools in which the fish are able to pass from one pool to another through a continuous flow of water, which maintains the water level in each of the pools.

The flow into and out of each pool is constant such that the level in each pool is constant. The static head is therefore equal to the velocity head (potential energy is equal to the kinetic energy). The fish are required to use bursts of speed to overcome the flow of water from one pool to the next. Burst speeds can be maintained for up to 20 seconds before resulting in exhaustion. Speeds are dependent on fish species, size, physical condition, and phase of life during migration as well as water quality. For each step,

$$h = \frac{U^2}{2g} = \frac{3^3}{2 \times 9.81} = 0.46 \text{ m} \tag{1.18}$$

The total number of steps for an overall elevation of 16 m is therefore 35 pools. This corresponds to 34 separate pools, since the first step is the upper river level. In fact, each pool has a submerged 1 m diameter opening, and the total distance of the ladder is 310 m. There are three additional larger pools that allow the fish to rest during their ascent.

What Should We Look Out For?

The velocity of the water from one step to another is determined by the level of the water from the step above as a simple conversion of potential energy to kinetic energy. In the Bernoulli equation expressed in head terms rather than energy, this is the conversion of pressure head to velocity head as given in Equation 1.18.

What Else Is Interesting?

There are a number of other well known designs of fish ladders that can overcome man-made obstacles in rivers, and include baffle fishways, fish elevators, and rock-ramp fishways. For example, the fish ladder at Hell's Gate on the Fraser River in British Columbia consists of double vertical-slot passages. This is similar to the pool-and-weir system, except that each "dam" has a narrow slot near the channel wall. This allows fish to swim upstream without leaping over an obstacle. Vertical-slot fish passages also tend to handle reasonably well the seasonal fluctuation in water levels on each side of the barrier.

Problem 1.7: Vessel Sizing and Testing

A vessel used to contain a liquid takes the form of a cylinder with radius r and height h with a hemispherical base (Figure 1.9). Given that the volume of the vessel is 45π, determine the radius of the vessel if its total surface area is to be a minimum.

Solution

The volume of the vessel comprises the cylindrical section and hemispherical end. That is,

$$45\pi = h\pi r^2 + \frac{2}{3}\pi r^3 \tag{1.19}$$

The height of the cylindrical section is, therefore,

$$h = \frac{45}{r^2} - \frac{2}{3}r \tag{1.20}$$

The area of the vessel comprises the cylindrical section and hemispherical end and also includes the top plate:

$$A = 2\pi rh + 2\pi r^2 + \pi r^2 \tag{1.21}$$

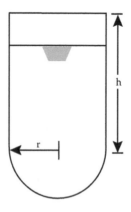

FIGURE 1.9
Cylindrical Vessel with Hemispherical Base

Using Equation 1.20, the area is

$$A = \frac{90\pi}{r} + \frac{5}{3}\pi r^2 \tag{1.22}$$

The radius for a vessel with the minimum surface area is found by differentiating

$$\frac{dA}{dr} = \frac{-90\pi}{r^2} + \frac{10}{3}\pi r \tag{1.23}$$

The minimum area with radius is found from

$$\frac{dA}{dr} = 0 \tag{1.24}$$

Solving gives a radius of 3 m.

What Should We Look Out For?

It is often the case that vessels used to contain fluids require a minimum of area to volume. This not only reduces heat loss or gain but also corresponds to the least amount of materials, such as metal, needed to fabricate the vessel and therefore minimise the costs.

What Else Is Interesting?

Vessels that have curved or hemispherical ends are often used to contain process liquid and gases that are either stored or operated under elevated pressure. The curved ends ensure a distribution of stresses to avoid vessel failure. To ensure that the vessels are capable of withstanding the pressure during commissioning, where appropriate, they may be pressure tested by filling with water. This is a form of hydrostatic pressure testing. It is a safe method as water is virtually incompressible, and by using colouring, leaks can be easily detected. It is also used to test pipework.

Problem 1.8: Air-Lift

Water is lifted from a large open vessel using a pipe with an internal diameter of 200 mm, which extends to a depth of 10 m in which compressed air is

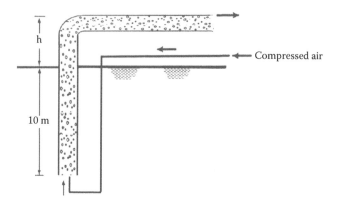

FIGURE 1.10
Air-Lift

admitted forming small bubbles forming a two-phase mixture with a combined average specific gravity of 0.8 shown in Figure 1.10. If the pipe extends to an elevation of 2 m above the surface of the water, determine the rate of flow of water.

Solution

An air-lift is a pumping device used to raise a liquid from a depth such as a well. It consists of a vertical pipe or leg extending down into the liquid such as a well into which compressed gas such as air is injected into the bottom. The bubbly mixture reduces the apparent density of the mixture in the pipe, and as the bubbles rise, the reduced hydrostatic pressure in the leg, which is unequal with the surrounding body of liquid, results in liquid being drawn into the leg. The air or gas is disengaged from the liquid at the top of the leg, which would correspond to an equilibrium elevation above the surface. Instead, the bubbly mixture is drawn off below this equilibrium elevation. The air-lift is used for raising oil from wells. In practice, the air or some other gas is admitted either continuously or intermittently and depends on the well geometry and properties of the oil. The optimal amount is usually based on well tests.

Ignoring frictional effects along the inside of the pipe and effects into and out of the pipe, the rate of flow of water drawn to the surface at an elevation of 2 m is found by applying Bernoulli's equation:

$$H = \frac{\rho_m g(H + h)}{\rho g} + \frac{U^2}{2g} \tag{1.25}$$

The velocity of the air/water mixture is therefore,

$$H = \sqrt{2g\left(\frac{10\rho_m - 2\rho)}{\rho}\right)} = \sqrt{2g\left(\frac{10 \times 1000 - 2 \times 800)}{1000}\right)} = 2.8\,\mathrm{ms}^{-1} \qquad (1.26)$$

The mass flow rate is

$$\dot{m} = \rho_m \frac{\pi d^2}{4} U = 800 \times \frac{\pi \times 0.2^2}{4} \times 2.8 = 70.3\ \mathrm{kgs}^{-1} \qquad (1.27)$$

The equilibrium elevation at which there will be no flow occurs at

$$\rho g H = \rho_m g(H + h) \qquad (1.28)$$

to give

$$h = H\left(\frac{\rho}{\rho_m} - 1\right) = 10 \times \left(\frac{1000}{800} - 1\right) = 2.5\ \mathrm{m} \qquad (1.29)$$

What Should We Look Out For?

If the leg is particularly deep, considerable air pressure will be required to overcome the hydrostatic pressure. Allowances should therefore be made in the calculation to allow for the compressibility of the injected gas. As the bubbles rise, expansion of the gas due to the progressive reduction of hydrostatic pressure causes an increase in upward velocity. If the bubbles coalesce, slugs may form which rise rapidly up the lift and can disengage at the surface alarmingly. Well known in the offshore oil industry, slug catchers are used to disengage gas from oil from mixtures raised from wells.

What Else Is Interesting?

Air-lift pumps are noted for their simplicity and low maintenance. They are an effective means of pumping oxygen-rich water in aquaponic applications in which plants in water are grown using water that has been used to cultivate aquatic organisms. Industrially, the use of compressed air as a source of oxygen within air-lift and bubble column reactors involves the process of sparging at the bottom as bubbles to promote oxygen transfer and cause circulation of the liquid. The reactor design is cylindrical and mounted on its axis and has an inner draught tube up which the air or oxygen rises. Alternatively, an external-loop air-lift-type reactor consists of a U-tube within which the sparging takes place, promoting oxygen transfer and liquid circulation. Compressed air sparged as bubbles within tubes has also been successfully used by divers to recover treasure from wrecks on the seabed where water with entrained sand particles are drawn up the tube. The sand and bubbles disengage at the top of the tube allowing the treasure to be uncovered without damage. A spectacular use of such a device is as a vent on Lake Nyos in Cameroon. Like a number of volcanic lakes, the lake is saturated at a depth with carbon dioxide and methane gas. In 1986, an overturn of the lake thought to be due to volcanic activity released the gases to the surface. The sudden and rapid release of a cloud of carbon dioxide which drifted onto land was responsible for killing around 1,700 people and 3,500 livestock. A vent pipe has subsequently been installed that penetrates down into the depths to allow the controlled de-gasing of the lake. The disengagement of combined water and gas at the lake surface gives rise to a spectacular plume some 50 m high.

Problem 1.9: Liquid–Liquid Separator

A mixture of oil and an aqueous solution is to be separated in the device shown in Figure 1.11. The aqueous solution has a density of 1100 kgm^{-3}, whereas the oil phase can range in density from 600 to 850 kgm^{-3}. The mixture is to be fed slowly to the separator in which the lighter oil phase will overflow while the

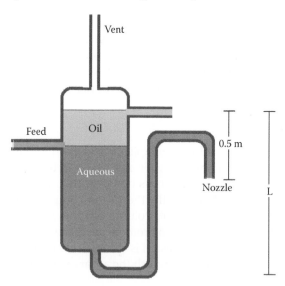

FIGURE 1.11
Water/Oil Separator

more dense aqueous phase will drip from the nozzle. The distance between the nozzle and the overflow is 50 cm. Determine the minimum distance L which can be used if the oil phase is not to drain through the nozzle.

Solution

Combining the hydrostatic legs of both the chamber and discharge leg gives the maximum available depth for the aqueous/organic phase boundary:

$$\rho_o g L = \rho_{aq} g (L - 0.5) \tag{1.30}$$

The minimum chamber length required corresponds to the greatest density of the oil phase (850 kgm^{-3}) that will ensure no loss of the oil phase with the aqueous phase. That is,

$$L = \frac{0.5\rho_{aq}}{\rho_{aq} - \rho_o} = \frac{0.5 \times 1100}{1100 - 850} = 2.2 \text{ m} \tag{1.31}$$

What Should We Look Out For?

The chamber is required to separate two immiscible phases. A good design of separator would allow the interface to be located at the feed point. There should also be a sufficient volume and cross-sectional area so as to allow sufficient time for the separation of dispersed phases, which are typically in the form of droplets.

What Else Is Interesting?

The dimensions of the chamber are such that the underflow of the oil phase will never leave by way of the nozzle. In this case, the depth will be greater than 2.2 m. If the chamber has a length of 3 m, say, the position of the interface as a function of density of the oil phase would still be

$$L = \frac{0.5\rho_{aq}}{\rho_{aq} - \rho_o} \tag{1.32}$$

Further Problems

1. Both legs of a U-tube contain water with a density of 1000 kgm^{-3}. If the left-hand leg also contains a layer of oil, which has a density of 800 kgm^{-3} to a depth of 10 cm, determine the difference in surface

levels of left- and right-hand legs if both legs are open to the atmosphere and the pressure difference needed to be applied between the
two legs if the surface levels are to remain the same. *Answer*: 2 cm,
196 Nm^{-2}

2. The double U-tube configuration as shown is used to measure the
density of a liquid whose density is less than that of water. Leaving
your answer in algebraic terms, determine the density of the
unknown fluid, ρ, in terms of the various column heights. *Answer*:

$$\rho = \rho_w \frac{h_1 - h_2 + h_3 - h_4}{h_3 - h_2}$$

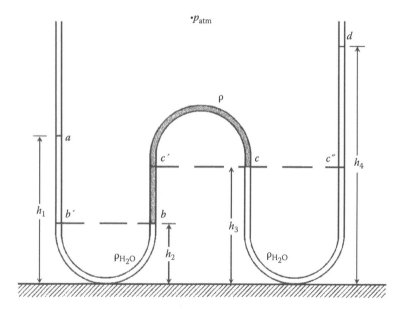

3. An inclined leg manometer is used to measure the pressure drop
across an air filter. The manometric fluid is a coloured oil with a
density of 800 kgm^{-3}. If the inclined leg is set at an angle of 30° to the
horizontal and the manometric fluid moves a distance of 15 cm along
the leg, determine the pressure drop across the filter. If the angle of
the inclined leg is adjusted to 20°, determine the magnification of
accuracy of the pressure drop measurement. *Answer*: 589 Nm^{-2}, 47%

4. Determine the gauge pressure of the water in the bulb. The densities of mercury and water are 13,600 kgm^{-3} and 1000 kgm^{-3}, respectively. *Answer:* 47.5 kNm^{-2}

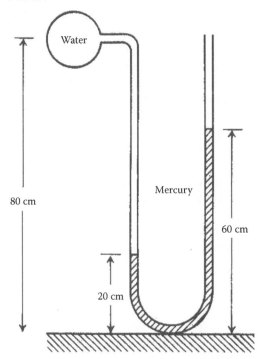

5. Two bulbs A and B contain air and are connected together as shown below. Determine the difference in pressure between the two bulbs.

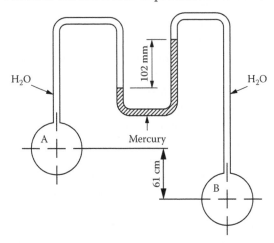

Answer: 6624 Nm^{-2}

6. Describe the operation of a differential manometer and provide an example of its type and use. Comment on the choice of manometric fluid.

7. Explain what is meant by gauge pressure and absolute pressure.

8. The drainage of liquid of density 995 kgm^{-3} from a vessel with a diameter of 5 m uses a siphon which consists of a flexible tube that extends down into the liquid and rises to an elevation of 3 m above the liquid level when full. If the liquid has a vapour pressure of 50 kNm^{-2}, determine the lowest discharge elevation below the highest point of the tube that avoids the formation of a vapour lock and the amount of liquid transferred. Standard atmospheric pressure is 101 kNm^{-2}. *Answer*: 5.25 m; 44.16 m^3

9. A rock sits in the bottom of a boat, which is floating on the surface of a pond. If the rock is thrown overboard into the pond, determine whether the level of the pond would rise, fall, or remain unchanged. *Answer*: Fall

10. Together with a clearly labelled sketch, show that the differential pressure Δp in an inclined leg manometer may be given by

$$\Delta p = \rho g L \left(\frac{a}{A} + \sin \theta \right)$$

where ρ is the density of the liquid, g is the gravitational acceleration, L is the distance along the included leg that the liquid moves, a is the cross-sectional area of the included leg, A is the cross-sectional area of the sump, and θ is the angle of inclination.

11. A hydrometer floats in water with 5 cm of its graduated stem unimmersed, and in oil of SG 0.7 with 3.5 cm of the stem unimmersed. Determine the length of stem unimmersed when the hydrometer is placed in a liquid of SG 0.9. *Answer*: 4.61 cm

12. A string supports a solid iron object of mass 180 g totally immersed in a liquid with a density of 800 kgm^{-3}. Calculate the tension of the string if the density of iron is 8000 kgm^{-3}. *Answer*: 1.59 N

13. A U-tube has a left-hand leg with a diameter of 5 cm and a right-hand leg with a diameter of 1 cm and inclined at an angle of 30°. If the manometric fluid is oil with a density of 920 kgm^{-3} and a pressure of 1 kPa is applied to the left-hand leg, determine the length by which the oil will have moved along the right-hand leg. *Answer*: 20.5 cm

14. An inclined alcohol-filled manometer is used to measure small pressure changes of a gas in a closed system. The manometer has a sump area of 16 cm^2 with a leg of cross-sectional area 0.25 cm^2 and overall length of 15 cm when open to atmosphere, included at an angle of 24° to the horizontal. If a distance of 12 cm is recorded

along the leg by the manometer fluid, determine the gauge pressure of the system. The density of alcohol is 780 kgm^{-3}. *Answer*: 388 Nm^{-2}

15. A U-tube has a left-hand leg with a diameter of 5 cm and a right-hand leg with a diameter of 1 cm and inclined at an angle of 24°. If the manometer fluid is oil with a density of 920 kgm^{-3} and a pressure of 400 Nm^{-2} is applied to the left-hand leg, determine the length by which the oil will have moved along the right-hand leg. *Answer*: 9.9 cm

16. An oil with a density of 880 kgm^{-3} and viscosity of 1 mNsm^{-2} flows through a horizontal 45° elbow with a uniform inside diameter of 10 cm with a mean velocity of 1 ms^{-1} and gauge pressure of 0.5 bar. Determine the magnitude and direction of the horizontal force exerted by the oil on the bend. *Answer*: 310 N, 22.5°

2

Flow Measurement

Introduction

Virtually all systems and processes involving the flow of fluids require the careful, accurate, and precise measurement of flow. This is essential not only to ensure required operational conditions are met and maintained but also to ensure safe conditions are not breached. Accurate measurement of flow is also required for accountancy purposes such as for the flow of oil and gas from wells.

The devices used to measure the flow of fluids are known as *flowmeters*. It is important to select correctly the most appropriate flowmeter for a particular application. This requires a sound knowledge and understanding of not only the behaviour of the flowing fluid properties but also an understanding of the operating principles of the flow measurement devices. Such flowmeters are broadly classified into those that are intrusive and those that are nonintrusive to the flow of the fluid. Broadly, they include differential pressure meters, positive displacement meters, mechanical, acoustic, and electrically heated meters. In general, most measure the flow of single-phase fluids. In the case of more complex multi-phase flow mixtures, such as combined oil and gas flows from offshore reservoirs, the measurement of both phases involves separation, conveying and measuring them separately as they flow through pipelines along the seabed.

A number of widely used flowmeters are based on the principle of converting one energy form to another, which can be measured. Examples include the orifice, nozzle, and Venturi meters that operate by constricting the flow area causing an increase in the velocity with a corresponding decrease in pressure. A measure of the pressure drop across the constriction gives an indirect measure of the rate of flow. Conversely, the variable area flowmeter operates with a float contained within a tapered tube in which the pressure drop across the float is fixed while the area around the float is variable. The elevation of the float gives an indication of the rate of flow. A calibration scale on the tapered tube is often used to account for the choice of float and properties of the fluid, including possible density variations due to temperature.

Other popular differential flowmeters include Dall tubes, Pitot tubes, gate meters, and target flowmeters. Many other types of commonly used flowmeters include turbine meters and positive displacement flowmeters including rotary and reciprocating piston, and sliding-vane-type flowmeters. Rotary piston flowmeters, for example, are commonly used in the water industry and in domestic metering and operate with a cylindrical working chamber that houses a hollow cylindrical piston of equal length. The central hub of the piston is guided in a circular motion by two short inner cylinders. The piston and cylinder are alternately filled and emptied by the fluid passing through the meter. The rotary movement of the piston is transmitted via a permanent magnet coupling from the drive shaft to a mechanical register. With a relatively low cost, they provide accuracies of around 2%, although they do not function well with viscous fluids.

Turbine meters are used in large pipeline installations but are less accurate than displacement flowmeters, particularly at low flow rates. They operate using a freely rotating propeller with blades mounted on a rotating shaft. Flow is generally straight through the meter, thereby permitting higher flow rates and less pressure loss than positive-displacement-type meters. As the rotor rotates, electrical pulses are generated, which are in proportion to the total volume that has passed through the meter. To protect the meter from harmful and abrasive solid particles, strainers are usually required to be installed upstream of the meter.

In the case of all flowmeters, the presence of particles, including solids and gas bubbles such as the result of cavitation and deaeration, are responsible for causing erroneous and inaccurate readings. The phenomenon of cavitation occurs inevitably as a result of insufficient pressure at the meter and can usually be prevented by installing a back-pressure valve downstream of the flowmeter. This will ensure that pressure within the meter never falls below the minimum value stipulated by the manufacturer of the device.

Cavitation is responsible for causing significant and substantial metering errors and inaccuracies. Where this occurs in oils, fuels, and solvents, cavitation releases a cloud of bubbles that may persist for considerable periods of time. The most notorious offenders are flow control valves, which are liable to cavitate whenever they are left in the nearly closed position, and certain types of three- and four-way valves. These tend to produce a burst of cavitation whenever they are changing position.

A number of methods have been devised to detect the phase velocities of particles in fluids. Optical methods are not always useful, however, due to distributed conditions caused by cavitation and deaeration. Nonintrusive metering methods are preferred due to the damaging and abrasive effect of the particles.

The entrainment of air in flowmeters can also lead to errors in liquid flowmetering and is to be avoided. Air entrainment generally results in air pockets of a larger size than the fine bubbles produced by cavitation. Large pockets of air, however, are relatively easy to remove from a flowing liquid before it enters a flowmeter by bleeding a line and in some cases is often a cheaper alternative than preventing the air from being entrained in the first place. It may be that after a piping system has been shut down and drained of liquid for maintenance, it can be difficult to fill again. Pockets of air are therefore liable to lodge in all the high points in the system, such as the tops of inverted U-bends, only to break up and pass through the system at some later point when the flow rate or pressure may fluctuate suddenly. Such high points can be kept to a minimum by good design or, where it is necessary, are supplied with an air bleed. These can be vented to allow trapped air at the time of filling to be released. For piping systems that feature many high points, it may be beneficial to install a permanent air separator upstream of the flowmeter.

It is surprisingly easy for air to be sucked into a pipe at a joint where the seal is only slightly defective. Since air is considerably less viscous than any liquid, seals that are adequate for keeping liquid inside a pipe are quite inadequate for keeping out air. It is not only the joints in suction lines that are vulnerable, but the joints in pipes which are just above atmospheric pressure can also allow air to be drawn in where pulsations may produce momentary subatmospheric pressures in the line. It is good practice in metering systems where there is no air separator for all pipe joints in suction lines to be inspected occasionally and for their seals to be replaced at the first sign of damage.

Air or gas separation devices are often confined to certain applications, usually on account of the value of the fluids being measured. Their use is obligatory, however, where valuable liquids, and especially petroleum products, are being metered intermittently for the purpose of trade or taxation. Such applications include petrol pumps, metering systems on road tankers, and metering systems at depots where delivery vehicles are loaded. Such devices usually consist of a vessel, often with some internal baffles added, which reduce the flowing liquid and give the entrained air plenty of opportunity to rise to and disengage from the top of the vessel. The air is vented through a valve controlled by a float, which prevents any liquid from escaping. Other types of gas separators utilise centrifugal force to expedite the passage of air through the liquid. Gas separators are heavy-duty devices designed to deal effectively with all forms of air entrainment and are capable of removing a proportion of cavitation bubbles. The gas extractor is a simple and inexpensive device, which is effectively a smaller version of the gas separator but without the internal baffles. It is not capable of removing the large quantities of air that arise through vortex entrainment, especially if this air happens to occur in the form of small bubbles rather than large gulps. The gas extractor is intended solely for the much lighter duty of removing pockets of air formed during filling of the system or as a result of contraction during cooling. It should never be used in systems where either vortex entrainment or cavitation might occur.

Problem 2.1: Venturi Meter Calibration

A Venturi meter located in a horizontal pipe with an internal diameter of 50 mm is used to measure the rate of flow of water. A calibration is provided for the flowmeter and is given by $\dot{Q} = 0.096\sqrt{h}$ where the flow rate is expressed in litres per second and the manometric head of a U-tube manometer containing mercury manometer is expressed in millimetres. Determine the assumed coefficient of discharge if the throat of the Venturi has a diameter of 16 mm.

Solution

A Venturi meter is a device used to measure the rate of flow of a fluid and consists of a tapering cone section that channels the flow into a narrow throat section that increases the velocity for which there is a corresponding reduction in pressure, known as the *Venturi effect*. Named for the Italian physicist Giovanni Battista Venturi (1746–1822), the differential pressure produced by the flowing fluid through the throat gives a measure of the rate of flow. The pressure differential can be measured by pressure gauges, transducers, or devices such as differential manometers attached both upstream and at the throat. In practice, the theoretical rate of flow through the device is not actually achieved as friction effects within the device are ignored. To allow for this difference, a correction factor known as a *coefficient of discharge* is used. For a well-designed Venturi, the coefficient typically lies between a value of 0.95 and 0.98. It cannot exceed a value of 1.0.

The general equation for the volumetric flow rate of an incompressible fluid through a horizontal Venturi meter shown in Figure 2.1 is given by

$$\dot{Q} = C_d a_1 \sqrt{\frac{2\Delta p}{\rho\left(\left(\dfrac{d_1}{d_2}\right)^4 - 1\right)}} \tag{2.1}$$

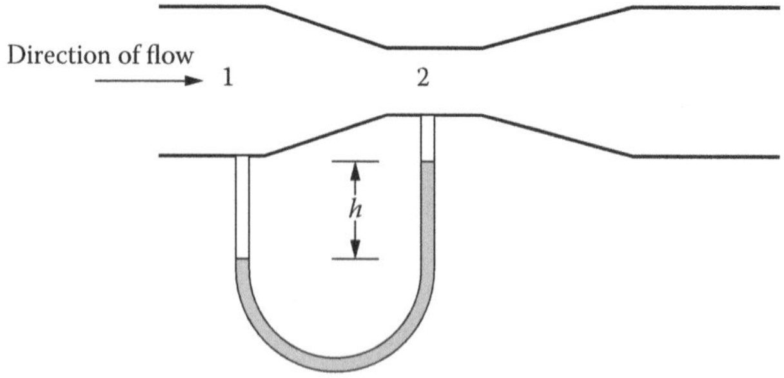

Direction of flow

FIGURE 2.1
Horizontal Venturi

For a mercury-filled U-tube differential manometer in which the difference in level is measured in millimetres, the measured pressure difference is

$$\Delta p = \left(\rho_{Hg} - \rho_W\right)gh = \left(13600 - 1000\right) \times 9.81 \times h / 1000 = 123.6h \text{ Nm}^{-2} \quad (2.2)$$

The supplied calibration formula is therefore

$$\frac{\dot{Q}}{1000} = C_d \frac{\pi \times 0.05^2}{4} \sqrt{\frac{2 \times 123.6 \times h}{1000\left(\left(\frac{0.05}{0.016}\right)^4 - 1\right)}} = 0.1 C_d \sqrt{h} \quad (2.3)$$

Relating this to the supplied calibration equation, the corresponding coefficient of discharge is

$$C_d = \frac{0.096}{0.1} = 0.96 \quad (2.4)$$

What Should We Look Out For?

The flow equation (Equation 2.1) is readily obtained by applying the Bernoulli equation between the throat and upstream position, together with the conservation of mass equation. That is, the flow through the meter remains unchanged irrespective of location. The equation can be modified for the flow of incompressible fluids in vertical or inclined pipes and for fluids that are significantly compressible. As a rule of thumb, a gas may be conveniently considered as being incompressible for differential pressures below 7 bar. Above this pressure, modifications for gas compressibility must be made to the equation for flow.

A well-designed Venturi allows good energy recovery. The dimensions of the Venturi are therefore important in minimising the permanent energy losses. Ideally, the Venturi consists of a rapid contraction and gentle expansion for which there are recognised and recommended standards of design. The use of computational fluid dynamics (CFD) permits a thorough analysis of the performance of the Venturi. Figure 2.2, for example, indicates the variation of velocity through the flowmeter, it can be seen that at the points

Flow

FIGURE 2.2
Velocity Profile in a Venturi Meter

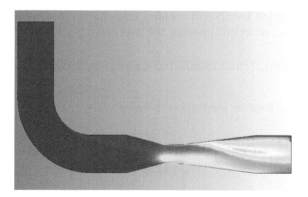

FIGURE 2.3
Location of a Venturi to a Bend

where the velocity is particularly high at the sharp corners and in the expansion cone, the velocity reduces again. There is a corresponding low pressure at these points.

Care should be taken in locating any flowmeter within a pipeline. It is recommended that a Venturi meter be installed in a straight length of pipe downstream with a distance of at least 10 to 20 pipe diameters away from any bends or other features that may be present. There should also be at least a further 5 pipe diameters of straight pipe beyond the flowmeter. Figure 2.3 illustrates a CFD simulation of the effects of the proximity of a Venturi to a bend. On the other hand, in view of the mixing pattern caused, they can be deliberately located near the bends to function effectively as inline mixers.

What Else Is Interesting?

The Venturi effect is the phenomenon in which a fluid flowing through a restriction increases in velocity with a corresponding decrease in pressure. An example of the Venturi effect is the wing on an airplane. As air passes over the curved upper portion of the wing, it must either compress against the air above, or increase in velocity. As the velocity rises, the pressure drops, thereby creating lift from the underside. A Venturi can also be used as a suction device to draw liquid from a reservoir such as for spraying paint (Figure 2.4).

Problem 2.2: Orifice Plate Meter

The rate of flow of water in a horizontal pipeline is measured using an orifice plate meter with a coefficient of discharge of 0.6. The pipeline has an inside diameter of 50 mm, and the orifice plate has a diameter of 35 mm.

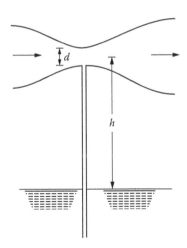

FIGURE 2.4
Venturi Lift

The pressure drop across the orifice is measured using a U-tube manometer containing mercury. If the difference in levels of the legs of the manometer is 100 mm, determine the mass flow rate of the water.

Solution

Orifice meters are a form of constriction across the flow of a fluid in a pipe. The placing of such constrictions in pipelines is a common way to measure the rate of flow of process fluids. The measurement of flow is by way of measuring a difference in pressure caused by the fluid increasing its velocity through the constriction and is therefore not a direct measure of flow. The pressure difference can be measured using pressure gauges or U-tube manometers from which the rate of flow can be conferred.

Orifice meters are low-cost meters noted for their simplicity with no moving parts and their accuracy. They are straightforward to install and maintain and are capable of covering a wide range of flow rates (see Figure 2.5). They are, however, an intrusive form of measurement and a flow constriction results in large permanent energy losses. Depending on the application, the orifice itself can become eroded, for example, by sand or corrosive fluids or may be obstructed by solids such as wax or hydrate in the case of hydrocarbon flow applications.

Using a fixed aperture for the measuring of differential pressure, the rate of flow can be calculated. The converse is the variable area flowmeter or rotameter that consists of a tapered tube within which a float is suspended by the upward flow of fluids. There is a fixed pressure drop across the float and variable area around the float in the tapered tube, which varies with elevation.

The rate of flow through an orifice in a horizontal pipe is determined by applying the Bernoulli equation at an upstream position and at the orifice

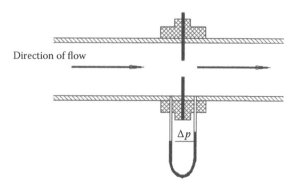

FIGURE 2.5
Orifice Plate Meter

itself. There are fixed and nonrecoverable energy losses from the flow through the orifice plate, and these are reflected in the coefficient of discharge, C_d. This is effectively the ratio of the actual to the theoretical rate of flow through the orifice. The value varies with the size and shape of the aperture and also the rate of flow through the orifice. Generally, the value approaches a value of 0.6 for Reynolds numbers measured through the orifice for values in excess of 10^4. The mass flow rate is given by

$$\dot{m} = C_d a \sqrt{\frac{2\rho \Delta p}{\beta^4 - 1}} \tag{2.5}$$

where β is the ratio of the orifice to pipe diameter. For the manometer, the pressure drop is determined from the difference in elevations of mercury in the two legs:

$$\Delta p = \left(\rho_{Hg} - \rho_w\right) gh = (13600 - 1000) \times 9.81 \times 0.1 = 12.36 \, \text{kNm}^{-2} \tag{2.6}$$

The mass flow is

$$\dot{m} = 0.6 \times \frac{\pi \times 0.05^2}{4} \times \sqrt{\frac{2 \times 1000 \times 12,360}{\left(\dfrac{0.05}{0.035}\right)^4 - 1}} = 9.19 \, kgs^{-1} \tag{2.7}$$

What Should We Look Out For?

The lower pressure occurs at the constriction caused by the increase the velocity. For the manometer, the location of the manometer would appear to be independent of the position relative to the pipe and flow. However, in practice, this is not the case as it is important to provide sufficient leg height to allow movement of the manometric fluid without reaching the tapping ports. The ports are located both upstream and downstream of the orifice plate, for which there are several recommended locations that range from corner tapping points to a more usual half a pipe diameter downstream and a whole pipe diameter upstream giving good measurements.

What Else Is Interesting?

Orifice meters are also used for compressible gas flow measurements. Determining the rate of flow requires good knowledge of the compressibility of the gas being measured for which the volumetric rate of flow can be determined from the following modified equation:

$$\dot{Q} = C_d a \sqrt{\frac{2\Delta p Z R T}{\left(\beta^4 - 1\right) p M_W}} \qquad (2.8)$$

The flow rate is therefore a function of the gas compressibility factor, Z. It is imperative that good compressibility data are known for the gas since errors in Z, particularly under high pressure flows, result in errors in the measured flow. A limitation of the flow of gases, however, is known as *choked flow*. This is a condition in which the fluid becomes limited in its flow or "choked" and is unable to be increased further. As the fluid passes through the orifice, a point is reached in which the rate of flow is unable to result in any further decrease in pressure, thereby limiting flow. The choking of gases occurs when the velocity leaving the orifice approaches sonic velocity—that is, at a Mach number of one. This results in shock waves that restrict flow causing the choking effect. The deliberate choking of gases can, however, be useful for limiting the rate of flow to processes. For liquids, the decrease in pressure below the vapour pressure results in partial flashing and cavitation, with the formation of vapour effectively limiting flow.

Problem 2.3: Evaluation of the Coefficient of Discharge

Following the installation and commissioning of an orifice plate meter in a pipeline used to determine the flow rate of cooling water to an industrial process, the meter was calibrated. This involved the collection of volumes of water, which was timed using a stopwatch whilst noting the corresponding pressure drop across the flowmeter. From the data presented below, determine the coefficient of discharge for the flowmeter when the theoretical flow rate, measured in litres per second, for the flowmeter is related to pressure drop, Δp (kPa) by $\dot{Q} = 0.18\sqrt{\Delta p}$.

Test number	1	2	3	4
Volume collected (litres)	31	58	33	99
Time(s)	60	90	45	120
Pressure drop (kPa)	20	30	40	50

Solution

Installing an orifice plate consists of inserting the plate, which is usually centrally drilled although other designs are possible, between the flanges at the ends of two lengths of pipe. Gaskets are used on either side to provide a good seal and prevent leakage. Commissioning is the final and thorough check.

For the orifice plate meter, the coefficient of discharge is an indication of the recovery of energy. It is expressed as the ratio of the actual flow rate of a fluid to the theoretical flow rate through an opening or constriction. When there is full recovery and no permanent energy loss, the coefficient of discharge has a value of 1.0. In this case, the actual flow rates are determined from the volume of water collected over measured time intervals and expressed as a ratio of the theoretical flow rates from the supplied equation. The coefficient of discharge values are presented in the table below.

Test number	1	2	3	4
Actual flow rate (litres s^{-1})	0.516	0.644	0.733	0.825
Theoretical flow rate (litres s^{-1})	0.805	0.986	1.138	1.273
Coefficient of discharge C_d	0.642	0.653	0.644	0.648

What Should We Look Out For?

The values for the coefficient of discharge have a mean value of 0.647. This is typical of orifice plate meters, but nonetheless, the range of values may have significance in terms of the accuracy of flow. In this case the standard deviation of the values is 0.0045. The accuracy of the flowmeter is the measure of the closeness or agreement of a numerical value to a true value. It is expressed as either significant figures or decimal places depending on whether proportional or absolute accuracy is important. For example, the value 0.0045 assumes that the four figures are meaningful. It would be incorrect to write the number to a precision of five significant figures unless the error in the estimate is indicated, such as 0.00450 ± 0.00005. In contrast, precision is a measure of the exactness of a measured quantity. The precision may be increased by increasing the sample size. In this case, more readings would need to be taken.

What Else Is Interesting?

While the coefficient of discharge for a well-designed Venturi meter is in the order of 0.95 to 0.98 (see Problem 2.1) signifying very good energy recovery, the coefficient for an orifice plate meter is much lower with poorer energy recovery. Orifice plate meters, however, are considerably cheaper than Venturi meters to fabricate and install and, consequently, are widely used. They are, however, not recommended for long-term use.

Problem 2.4: Pitot Tube Traverse

A Pitot tube is used to determine the volumetric flow rate of a gas in a duct of circular cross section with a diameter of 2.0 m. A Pitot traverse using nine data readings is used, as presented in the following table. Determine the rate of flow of the gas which has a density of 1.2 kgm^{-3}.

Position	1	2	3	4	5	6	7	8	9
Distance (cm)	80	60	40	20	0	20	40	60	80
Pressure (Nm^{-2})	126	220	261	278	284	280	269	252	175

Solution

The Pitot tube (Figure 2.6) is an instrument used to measure the velocity of a flowing fluid by measuring the difference between the impact pressure and static pressure in the fluid. The device normally consists of two concentric tubes; one with a face facing the flow to measure the impact pressure, the other with holes drilled perpendicular to the flow to measure the static pressure. By taking a number of readings at various points in the cross section of a pipe or duct, known as a Pitot traverse, the overall rate of flow can be determined from the velocity profile.

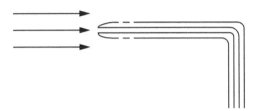

FIGURE 2.6
Pitot Tube

The local velocity in the duct is determined from

$$U = \sqrt{\frac{2\Delta p}{\rho}} \tag{2.9}$$

Position	1	2	3	4	5	6	7	8	9
Velocity (ms^{-1})	14.49	19.18	20.88	21.52	21.77	21.60	21.17	20.49	17.09

The average velocity across the cross section can be approximated to

$$\bar{U} = \frac{\sum U\Delta d}{d} \tag{2.10}$$

The overall flow rate can be found using numerical integration, which is a mathematical procedure used to calculate the approximate value of an integral. Such a technique is often used when a function is known only as a set of variables for corresponding values of a variable and not as a general formula that can be integrated. Simpson's rule and the trapezium rule are examples of numerical integration. In this case, the average velocity is 18.23 ms^{-1}. The volumetric flow rate is therefore

$$\dot{Q} = a\bar{U} = \frac{\pi \times 2^2}{4} \times 18.23 = 57.24 \text{ m}^3\text{s}^{-1} \tag{2.11}$$

It is assumed that the velocity at the wall is stationary.

What Should We Look Out For?

To get an accurate flow rate, readings should be repeated not only to provide a degree of statistical variation, if needed, but also to increase, when possible, the number of points on the traverse. These points can be located at defined points, but it is important then to determine the elemental flow for each point such that the total flow of their sum is determined as

$$\dot{Q} = \sum_{i=1}^{i} a_i U_i \tag{2.12}$$

For ducts of square or rectangular cross section, the area can be divided into smaller areas and the point velocities measured. Again, the total flow is the sum of the elemental flows in each area.

While the Pitot tube is usually slim in design, it is important that in their use within narrow pipes or tubes they do not provide an unnecessary obstruction. This would otherwise influence the velocity around the device and not give a true indication of the rate of flow.

FIGURE 2.7
Pitot Tube on the Nose of a Formula 1 Car (Photo from C.J. Schaschke.)

What Else Is Interesting?

As with all flow measurement devices, the Pitot tube should be ideally located away from disturbances such as bends. The device was devised by the Italian-born French engineer Henri de Pitot (1695–1771). Commercial and military aircraft and Formula 1 motor racing cars (Figure 2.7) use Pitot tubes to measure the velocity of the vehicles. A differential pressure transducer is used to provide an electronic signal to indicate velocity.

Problem 2.5: Venturi Flume

Water flows along a flume at a depth of 2 m. The flume features a 10-cm-high obstruction that causes the surface of the water to dip by 8 cm. Determine the rate of flow of the water per unit width of flume.

Solution

The Venturi flume (Figure 2.8) is a commonly used method of measuring the rate of flow in open channels and flume. The use of an obstruction at the floor causes a rise of the water over it, which reduces the flow area and increases the velocity. According to the Venturi principle, there is a corresponding decrease in pressure at this point. This is manifested as a decrease in the level at the surface.

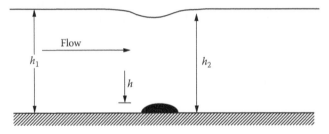

FIGURE 2.8
Venturi Flume

Applying the Bernoulli equation at an upstream location and at the obstruction itself gives an average velocity in the flume as

$$U_1 = \sqrt{\frac{2gh}{\left(\dfrac{h_1}{h_2}\right)^2 - 1}} = \sqrt{\frac{2 \times 9.81 \times 0.08}{\left(\dfrac{2}{1.82}\right)^2 - 1}} = 2.75 \text{ ms}^{-1} \qquad (2.13)$$

The rate of flow per unit width is therefore

$$\frac{\dot{Q}}{W} = U_1 h_1 = 2.75 \times 2 = 5.5 \text{ m}^3 \text{m}^{-1} \qquad (2.14)$$

What Should We Look Out For?

The datum elevation used in the Bernoulli equation can be applied at any depth, but this must be used consistently in the equation.

What Else Is Interesting?

The Parshall flume is the most commonly used type of Venturi flume. It is widely used to measure surface water and irrigation flow, as well as industrial discharges, municipal sewer flows, and flows to and from waste-water treatment plants. Named after Dr. Ralph Parshall of the U.S. Soil Conservation Service in 1915, the design consists of a uniformly converging upstream section with a short parallel throat section and a uniformly diverging downstream section. The base of the upstream section is flat and slopes downward in the throat before rising in the downstream section. The downstream elevation is below that of the upstream elevation.

Problem 2.6: Flowmeter Calibration by Dilution Method

A Venturi meter with a throat of 15 cm is used to measure the flow rate of water along a horizontal pipe that has an internal diameter of 17 cm. The flowmeter is calibrated using a solution of salt with a concentration of 10 gl^{-1} fed upstream of the meter at a rate of 50 millilitres per second. If the pressure difference between the upstream position and the throat is found to be 759 Nm^{-2} when a sample of salt-bearing water was analysed downstream from the flowmeter and found to be 15 mgl^{-1}, determine the flow rate of water through the pipeline and the coefficient of discharge of the flowmeter.

Solution

The precise and accurate calibration of measurement devices such as gauges and flowmeters is critical when they are used to control processes or have significance in terms of material accountancy. Prior to use, and at other times during routine operation, such devices should be calibrated or recalibrated.

The calibration of flowmeters is a method of adjusting an instrument so that its readings can be correlated to the actual values that are being measured. Calibration is achieved using methods such as recognised standards or by experimentation. In its simplest form, calibration of flowmeters can involve measuring known volumes through the meter and comparing the results to the measured values. Adjustments can be made to gauges, dials, calibration coefficients, or other variables to ensure that the meter accurately reflects reality and to check periodically for any drift.

The dilution method is a simple method for the calibration of flowmeters where other calibration methods may not be appropriate or practical. It involves the addition of an extraneous material at a known concentration. An amount is added either upstream or downstream of the flowmeter for which the mixed and diluted concentration is measured a distance further downstream. From a material balance, the actual flow can be determined and the flowmeter calibration completed.

In this case, salt added upstream is diluted such that the mass balance of the salt is

$$\dot{Q}_W C_W + \dot{Q}_S C_S = \left(\dot{Q}_W + \dot{Q}_S\right) C_o \tag{2.15}$$

That is, the flow of water is

$$\dot{Q}_W = \dot{Q}_S \left(\frac{C_o - C_S}{C_W - C_o}\right) = 50 \times 10^{-6} \times \left(\frac{0.0015 - 10}{0 - 0.0015}\right) = 0.0333 \text{ m}^3\text{s}^{-1} \tag{2.16}$$

For the Venturi member, the total rate of flow (water and salt solution which is added upstream) is given by

$$\dot{Q} = C_d a \sqrt{\frac{2\Delta p}{\rho(\beta^4 - 1)}} \tag{2.17}$$

The coefficient of discharge is

$$C_d = \frac{\left(\frac{\dot{Q}}{a}\right)^2}{\frac{2\Delta p}{\rho(\beta^4 - 1)}} = \frac{\left(\frac{0.0333 + 50 \times 10^{-6}}{\pi \times 0.017^2}\right)^2}{\frac{4}{2 \times 759}} = 0.92 \tag{2.18}$$
$$\frac{}{1000 \times \left(\left(\frac{0.017}{0.015}\right)^4 - 1\right)}$$

The coefficient of discharge is a little low for a Venturi meter (see Problem 2.1), and it would be worth carrying out further calibrations.

What Should We Look Out For?

It is possible to use the dilution method by adding the extraneous material downstream of the flowmeter itself. In this case, the flow through the flow-meter would therefore not include the extraneous material in the calculation. It is also possible to use salt as the extraneous material even if the process liquid should also contain salt.

What Else Is Interesting?

A calibration chart is a graphical representation of values read from an instrument. It is usually presented on the abscissa and the corrected value or quantity on the ordinate. An instrument should be calibrated with a sufficient number of data points to ensure a proper relationship between the measured and indicated values.

Further Problems

1. A Venturi is to be installed to measure the flow of water. The mass flow rate of water used in the process is not expected to exceed a rate of 5 kg⁻¹. The Venturi is to be located in a straight section of horizontal pipe with a 50 mm inside diameter and is to be connected to a mercury-filled differential manometer. If the maximum allowable pressure differential is 70 kNm⁻², determine the throat area required for the flowmeter. The coefficient of discharge may be assumed to be 0.96. *Answer:* 23 mm

2. Given that the mass flow rate equation for a rotameter (variable area flowmeter) is

$$\dot{m} = C_d a_o \sqrt{\frac{2 V_f \rho \left(\rho_f - \rho \right) g}{A_f}}$$

show that for a float height, h, in a rotameter tube of cone angle, α, this becomes

$$\dot{m} = C_d h \tan\left(\frac{\alpha}{2}\right) \sqrt{8\pi V_f \rho \left(\rho_f - \rho \right) g}$$

where m is the mass flow rate, a_o is the area of the annulus, V_f is the volume of the float, ρ is the density of the fluid, ρ_f is the density of the float, A_f is the impact area of the float, C_d is the coefficient of discharge, and g is the gravitational acceleration.

3. Determine the pressure differential recorded by two independent pressure gauges located at the two tapping points in a vertical venture for a flow of crude oil with a density of 850 kgm^{-3} which flows upward at a rate of 0.056 m^{3}s^{-1}. The Venturi meter, with a coefficient of discharge of 0.98, has an inlet diameter of 20 cm and a throat diameter of 10 cm. The vertical distance between the tapping points is 30 cm. *Answer*: 22.8 kNm^{-2}

4. Explain the various designs of float and the criteria used for selecting suitable floats and how an appropriate float density may be selected for a particular flow measurement application.

5. Discuss the merits and demerits of both Venturi and orifice plate meters as fluid flowmeters.

6. Starting with the Bernoulli equation, show that the volumetric rate of flow, $\dot{Q}$, of a process fluid through a horizontal Venturi meter can be given by

$$\dot{Q} = C_d a \sqrt{\frac{2\Delta p}{\rho(\beta^4 - 1)}}$$

where C_d is the coefficient of discharge, a is the internal cross-sectional area of the pipe, Δp is the pressure differential, ρ is the density of the process fluid, and β is the ratio of pipe to throat diameter.

7. Installed in a pipe with an inside diameter of 10 cm is a horizontal Venturi meter with a throat diameter of 5 cm, which is used to measure the flow of process cooling water to a furnace. During commissioning, the differential pressure measured between the throat and an upstream position is found to be 17.4 kPa for a rate of flow of 685 litres per minute. Determine the coefficient of discharge of the Venturi meter. *Answer*: 0.954

8. Showing relevant mass balance equations, describe the dilution method as a technique that can be used to calibrate flowmeters.

3

Freely Discharging Flow

Introduction

The storage and containment of a fluid is usually dependent on the form of the fluid, its properties, and its intended purpose. Small quantities of fluids such as water may be appropriately contained in bottles and jugs, whereas considerably larger volumes may be contained in reservoirs and lagoons. Gases may be contained in enclosed vessels to prevent leakage and escape or even underground caverns created naturally or artificially such as those formed from removing rock salt as brine. In some cases, gases may be liquefied such as liquid petroleum gas (LPG) for storage since liquids occupy a far smaller volume than in the gaseous state and therefore a smaller volume of vessel is conveniently required. They do require, however, a strong-walled vessel required to contain the pressure necessary to maintain the fluid in a liquefied state.

Tanks are a type of vessel used to contain liquids, usually at atmospheric pressure, and are used to store liquids until required for use, such as an oil tank. They may alternatively contain a liquid for an intended purpose, such as a bath, or be used to receive and contain a processed fluid, such as a septic tank. Tanks can also be used for transportation purposes by road or rail, such as delivery of petroleum and diesel fuels from a refinery to a service station for commercial retail.

Whatever the type of tank, vessel, or receptacle, the fluid of interest needs to be capable of being transferred, adequately contained, and discharged at the appropriate point. Filling vessels may be as simple as the action of pouring from one vessel into another. It is important to ensure that the receiving vessel has sufficient capacity and that there is adequate control to ensure that the receiving vessel does not overflow. The use of pipes with valves that regulate the direction and rate of flow is a more conventional approach for filling larger tanks. In the domestic lavatory cistern, the action of a float on the surface of the water is used to shut off the valve once a certain level has been attained to prevent overflow. Tankers that transfer fuel use flexible hoses and discharge controlled volumes. Dipsticks and sight glasses are simple methods of determining the filled depth of a liquid, such as oil in a lawnmower.

As with the filling of a vessel, discharging a vessel can be as simple as tilting the vessel to ensure the liquid overflows, such as a jug of milk. For larger vessels, tilting the vessel may not be practicable. A leg that penetrates down into the liquid may therefore be used, and either a suction applied to the leg can be used to lift the liquid out, or if practical, a positive pressure can be applied to the surface of the liquid to eject the liquid up through the leg. More usual is a pipe attached to the bottom or the side of the tank with a shut-off valve, in which the rate of discharge is influenced by the head pressure in the tank and the depth of the liquid above the valve. With the progressive loss of liquid, the level of the remaining liquid continues to influence the rate of discharge. This is readily observed with the drainage of a bath full of water through the plughole in which the level initially appears to fall rapidly but steadily reduces as the bath empties. Effectively converting potential energy of the water in the bath into kinetic energy through the plughole, the rate of discharge is proportional to the square root of the depth. The term *quasi-steady state* therefore refers to a system that is based on a steady-state model but is unsteady such that the depth changes with time.

Problem 3.1: Discharge through an Orifice

A very large open tank is used to store water (see Figure 3.1), and it is allowed to discharge through a circular, sharp-edged orifice with a diameter of 1.0 cm on the side of the tank. The level of water above the liquid is maintained at a constant 4 m. Determine the rate of discharge if the coefficient of discharge for the orifice is assumed to have a value of 0.6.

Solution

The tank has a large capacity such that it may be assumed that the drop in water level is negligible. The rate of discharge or outflow through the orifice is the same as the rate of inflow, thereby maintaining no level of change of the water in the tank. In addition to the depth of the water, the rate of discharge is dependent on the area of the orifice, the mean velocity, and the coefficient of discharge, which is used as an indication of the recovery of energy. Expressed as the ratio of the actual flow rate of a fluid to the theoretical flow rate through an opening or restriction, the coefficient has a value of 1.0 for full energy recovery and no permanent energy loss. For holes and orifices with sharp edges and high discharge flows, the value, however, may be as low as 0.6. (See Problem 2.2.)

The rate of flow is, therefore,

$$\dot{Q} = C_d a_o \sqrt{2gh} = 0.6 \times \frac{\pi \times 0.01^2}{4} \times \sqrt{2 \times 9.81 \times 4} = 4.17 \times 10^{-4}\,\mathrm{m^3 s^{-1}} \quad (3.1)$$

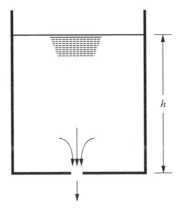

FIGURE 3.1
Tank Drainage

What Should We Look Out For?

The assumption that the tank has a large capacity is to assume that the level does not significantly change with time. The discharge is continuous, and would in fact decrease with the progressive loss of capacity and therefore head above the orifice. An example is the discharge from a reservoir (Figure 3.2).

Discharge calculations are made on the assumption that the liquid is able to discharge freely and without the complication of complex flow phenomena

FIGURE 3.2
Open Discharge of Laggan Dam, Scotland (Photo from C.J. Schaschke.)

such as the formation of a vortex. There is also the assumption that the properties of the liquid such as viscosity have no influence on the rate of discharge. There would evidently be a considerable difference between the discharge of viscous liquids such as honey, thick oils, and polymer solutions in comparison to water, light oils, and solvents. Further assumptions include the tank having a large cross-sectional area such that the walls of the tank have no influence and that the surface of the liquid in the tank is maintained at the same pressure as the discharging liquid. In situations where this is not the case, glugging occurs as air or gas is drawn back through the orifice in an attempt to reach an equilibrium pressure, which occurs when pouring a liquid such as beer through the narrow neck of a bottle.

What Else Is Interesting?

A vortex is a rapidly rotating fluid. It occurs in nature as whirlpools, maelstroms, and tornadoes. With discharging liquid from a tank, a free vortex is often formed in which the tangential velocity is proportional to the inverse of the distance from the axis. The streamlines move freely in horizontal concentric circles with no variation of the total energy across them. A forced vortex is also a rotating fluid in which the circular streamlines of liquid are caused by an external source such as a stirrer or impeller.

Problem 3.2: Reservoir Inflow

The capacity of a water reservoir is related to depth by the equation

$$V = \frac{\pi h^2}{12} \tag{3.2}$$

If the reservoir receives a flow of water at the constant rate of 2 m³ per minute, determine the rate at which the depth increases when the depth is already 4 m.

Solution

The change in capacity, or volume, can be found by differentiating Equation 3.2—that is,

$$dV = \frac{\pi h}{6} dh \tag{3.3}$$

The (constant) rate of flow into the reservoir is equal to the change in volume with time:

$$\dot{Q} = \frac{dV}{dt} = \frac{\pi h}{6}\frac{dh}{dt} \tag{3.4}$$

The change in depth with time is therefore,

$$\frac{dh}{dt} = \frac{6\dot{Q}}{\pi h} = \frac{6 \times 2}{\pi \times 4} = 0.955 \text{m.min}^{-1} \tag{3.5}$$

What Should We Look Out For?

In cases of simple mathematical relationships of geometry, analytical solutions are usually straightforward. Apart from the most straightforward forms of tank geometry, however, the relationship for reservoir volume is not likely to be known. For cases involving simultaneous inflow and outflow, such as through a pipe to feed a hydro-turbine, the problem becomes considerably more difficult to solve analytically.

What Else Is Interesting?

The use of the equation expressing the geometry of a reservoir or tank can be useful in determining the capacity which may be useful for safe operational control or accountancy purposes. For example, the time required to increase the volume from a depth of 6 m to 12 m can be readily calculated with a substitution of the depths into Equation 3.2 to give volumes of 9.42 m³ and 37.7 m³, for which the time taken is 14.14 minutes at a flow rate of 2 m³ per minute. Alternatively, the time can be found by integrating Equation 3.4:

$$\int_0^t dt = \frac{\pi}{6\dot{Q}} \int_{h_1}^{h_2} h\,dh \tag{3.6}$$

to give

$$t = \frac{\pi}{6\dot{Q}}\left[\frac{h^2}{2}\right]_6^{12} = \frac{\pi}{6 \times 2}\left[\frac{12^2 - 6^2}{2}\right] = 14.14 \text{min} \tag{3.7}$$

Problem 3.3: Laminar Flow

A honey production process involves the transfer of honey from a large feed vessel to a smaller receiving vessel under the influence of gravity by way of a connecting tube. The feed vessel has a large capacity, whereas the receiving vessel is considerably smaller with a uniform cross-sectional area of 0.1 m^2. The connecting tube is 5 m long with an internal diameter of 12 mm and enters the receiving vessel submerged. If the difference in levels between the two vessels is initially 1 m, determine the time taken for the difference in levels between the tanks to fall to 0.5 m. The viscosity of the honey is 1.0 Nsm^{-2}, and the density is 1360 kgm^{-3}.

Solution

Since the honey has a high viscosity, and noting that the tube has a narrow bore, the flow between the feed and receiving vessels may be assumed to be laminar. The rate of flow is determined by the difference in levels between the vessels and the frictional resistance within the tube. For laminar flow, the rate of the flow is determined from the Hagen-Poiseuille equation and is equal to the change in capacity of the smaller vessel—that is,

$$A\frac{-dh}{dt} = \frac{\pi\rho g h d^4}{128\mu L} \tag{3.8}$$

Rearranging and integrating from an initial difference in elevations of 1.0 m to a final difference of 0.5 m gives a time of

$$t = \frac{128\mu L A}{\pi\rho g d^4}\ln\left(\frac{h_1}{h_2}\right) = \frac{128\times1\times5\times0.1}{\pi\times1360\times9.81\times0.012^4}\ln\left(\frac{1}{0.5}\right) = 51067 \text{ s} \tag{3.9}$$

That is, it gives a time of over 14 hours.

What Should We Look Out For?

Honey is known to exhibit Newtonion flow characteristics for which the viscosity is notably dependent on temperature. In this case, the viscosity is relatively low for honey, but the important check is the Reynolds number which, in this case, confirms laminar flow and the use of the Hagen-Poiseuille equation. The greatest flow and therefore the greatest Reynolds number occur when the difference in levels is greatest:

$$\dot{Q} = \frac{\pi\rho g h d^4}{128\mu L} = \frac{\pi\times1360\times9.81\times1\times0.012^4}{128\times1\times5} = 1.36\times10^{-6} \text{ m}^3\text{s}^{-1} \tag{3.10}$$

and

$$Re = \frac{4\rho\dot{Q}}{\pi\mu d} = \frac{4\times1360\times1.36\times10^{-6}}{\pi\times1\times0.012} = 0.2 \qquad (3.11)$$

This thereby confirms laminar flow in a tube with a circular cross section.

What Else Is Interesting?

The equation for transfer suggests that the time taken to reach the same level between the two vessels would be an infinite period of time. It is more usual and practical to determine the time it takes to reach a certain proportion of the final difference.

Problem 3.4: Tank Drainage

An open cylindrical vessel mounted on its axis contains a liquid to a depth of 5 m and is allowed to discharge freely from a sharp-edged orifice with a diameter of 25 mm located in the side of the tank at a depth of 2 m. If the vessel has a diameter of 4 m, determine the depth remaining in the tank after 30 minutes. The coefficient of discharge of the orifice may be taken as 0.6.

Solution

The change in capacity of a vessel that discharges freely through an orifice can be determined from a quasi-steady state mass balance:

$$A_t\frac{-dh}{dt} = C_d a_o \sqrt{2gh} \qquad (3.12)$$

The depth after 30 minutes can be found by integration as

$$h_2 = \left(\frac{-tC_d a_o}{A_t}\sqrt{\frac{g}{2}} + \sqrt{h_1}\right)^2$$

$$= \left(\frac{-1800\times0.6\times0.025^2}{2^2}\times\sqrt{\frac{9.81}{2}} + \sqrt{5-2}\right)^2 = 1.84\text{ m}$$

$$(3.13)$$

The total depth is therefore $1.84 + 2 = 3.84$ m.

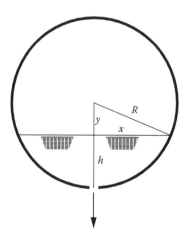

FIGURE 3.3
Geometry for Cylindrical and Spherical Tanks

What Should We Look Out For?

Where the cross section of the tank is not constant with the depth, such as a cylindrical vessel on its side or a spherical tank, the area of the tank in the calculation can no longer be considered to be constant but must be related to the depth (Figure 3.3). For these two cases, the area of the liquid is related to depth using the Pythagoras method—that is,

$$R^2 = x^2 + y^2 \tag{3.14}$$

and

$$R = y + h \tag{3.15}$$

For a cylindrical tank of length, L, the surface area of the tank is

$$A_t = 2xL = 2\sqrt{\left(2Rh - h^2\right)}L \tag{3.16}$$

while for a semi-spherical vessel of surface radius, x,

$$A_t = \pi x^2 = \pi\left(2Rh - h^2\right) \tag{3.17}$$

The associated integrations are then straightforward. Situations in which there is a simultaneous inflow with outflow are notably more complex. A tank will continue to increase (or decrease) in capacity until the outflow is balanced with the inflow, where the rate of outflow is related to the depth. That is,

$$A_t \frac{dh}{dt} = \dot{Q}_{in} - \dot{Q}_{out} \tag{3.18}$$

To solve this analytically requires a more complex mathematical integration. The equilibrium depth is where there is no change of depth with time such that the rate of flow out is equal to the rate of flow in. That is,

$$h = \left(\frac{\dot{Q}_{in}}{a\sqrt{2g}} \right)^2 \tag{3.19}$$

As an example, the flow into and out of dammed reservoirs is an important aspect of water management. Water flow can increase from runoff from tributaries and will increase due to rainfall. Level control is by way of weirs and sluices.

What Else Is Interesting?

This solution is based on the assumption that the liquid discharges freely through the orifice and that the level of the liquid falls uninterrupted. There is therefore no vortex, the viscosity of the liquid has no influence, the coefficient of discharge of the orifice is constant across all flow rates, and the tank is of sufficient width such that the wall of the tank plays no role. These assumptions are particularly general if one considers the likely difference between the free discharge of water compared to that of a thick and viscous oil.

Problem 3.5: Tank Drainage through a Connecting Pipe

Water flows from a cylindrical tank that is 1 m in diameter through a horizontal pipe has a 5 cm internal diameter and 35 m length (Figure 3.4). Determine the time it takes to lower the water surface in the tank from 3 m to 0.5 m above the open end of the pipe. Allow for entry and exit losses, and assume a friction factor as 0.005.

Solution

The losses along the pipe are due to frictional losses and the combined entry and exit losses which in head form are

$$h = \frac{4fL}{d} \frac{U^2}{2g} + 1.5 \frac{U^2}{2g} \tag{3.20}$$

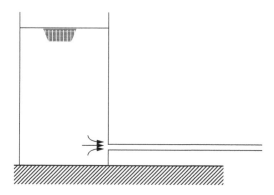

FIGURE 3.4
Tank Drainage through a Pipe

As the tank discharges, the change in capacity of the straight-walled tank such that only the depth changes with time is

$$aU = -A_t \frac{dh}{dt} \tag{3.21}$$

Combining Equations 3.20 and 3.21, the time taken to lower the level of liquid can therefore be found by integrating

$$\int_0^t dt = \frac{-A_t}{a\sqrt{\dfrac{2g}{\dfrac{4fL}{d}+1.5}}} \int_{h_1}^{h_2} h\, dh \tag{3.22}$$

to give

$$t = \frac{2A_t}{a\sqrt{\dfrac{2g}{\dfrac{4fL}{d}+1.5}}} \left(\sqrt{h_1}-\sqrt{h_2}\right)$$

$$= \frac{2 \times \dfrac{\pi \times 1^2}{4}}{\dfrac{\pi \times 0.05^2}{4}\sqrt{\dfrac{2 \times 9.81}{\dfrac{4 \times 0.005 \times 35}{0.05}+1.5}}} \times \left(\sqrt{3}-\sqrt{0.5}\right) = 729\ \text{s} \tag{3.23}$$

What Should We Look Out For?

This problem is considerably simplified by specifying the friction factor that remains constant throughout the draining process. In reality, however, the friction factor and, hence, rate of flow will vary as the tank discharges. While an analytical solution is not possible unless the variation of the friction factor with the flow is fully known, a computer iteration is the most convenient approach to take, particularly where a quick analytical answer is not required.

What Else Is Interesting?

As the tank discharges, the pressure drop along the pipe through which the liquid flows progressively decreases. However, if the end of the tank were to be blocked to stop the flow, the pressure at the end would be equal to the static pressure of the liquid in the tank. If the pipe is then instantaneously "unblocked," there would be a rapid surge in flow and a decline again to a rate where the forces due to the head in the tank and frictional resistance in the pipe return to an equilibrium.

Problem 3.6: Drainage between Tanks

A catchment reservoir beneath a small water cooling tower measures 4 m by 2 m in area. Prior to recirculation of the water for reuse, the water flows into another smaller reservoir with an area of 2 m by 2 m by way of a submerged opening with a flow area of 100 cm² with an assumed coefficient of discharge of 0.62. During normal operation the level in the larger reservoir is 30 cm above that in the smaller reservoir. If water cooling tower operation is halted with neither further addition of water to the catchment nor recirculation, determine the time it takes to reduce the difference in levels between the two catchments to 10 cm.

Solution

Under normal operation, the flow of one reservoir to the other ensures a constant difference in level. However, when the normal operation ceases, the larger reservoir will drain freely into the small reservoir. The level of the larger reservoir will therefore fall while the smaller will rise to eventually reach the same level (Figure 3.5). The change in the difference in levels between the two reservoirs, dh, is therefore

$$h - dh = h - dH_1 - dH_2 \qquad (3.24)$$

The changes in both levels are related by displacement from one reservoir to the other. That is,

$$dH_2 = \frac{A_1}{A_2} dH_1 \qquad (3.25)$$

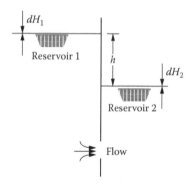

FIGURE 3.5
Flow between Reservoirs Connected by a Submerged Orifice

Combining Equations 3.24 and 3.25, gives

$$dh = dH_1 \left(1 + \frac{A_1}{A_2} \right) \tag{3.26}$$

The change in capacity is therefore

$$\frac{-A_1}{\left(1 + \dfrac{A_1}{A_2} \right)} \frac{dh}{dt} = C_d a \sqrt{2gh} \tag{3.27}$$

Integrating between the two depths over the time period, t, gives

$$t = \frac{2A_1 \left(\sqrt{h_1} - \sqrt{h_2} \right)}{C_d a \sqrt{2g} \left(1 + \dfrac{A_1}{A_2} \right)} = \frac{2 \times 4 \times 2 \times \left(\sqrt{0.3} - \sqrt{0.2} \right)}{0.62 \times 10^{-2} \times \sqrt{2 \times 9.81} \times \left(1 + \dfrac{4 \times 2}{2 \times 2} \right)} = 45 \text{ s} \tag{3.28}$$

What Should We Look Out For?

For a closed system in which there is neither flow in or out, the upper level freely discharges through the submerged orifice such that the level of the lower reservoir rises. The flow will continue until the two levels are the same, providing there is sufficient capacity in the smaller reservoir, otherwise the reservoir will overflow.

What Else Is Interesting?

A cooling tower is used to condense steam or reduce the temperature of water used as a cooling medium in a process for reuse. Cooling towers are either natural draught or forced draught in design and operation. Natural draft cooling towers are large structures that contain packing material with a high specific surface area down which the water to be cooled trickles and cascades, contacting with cool air which is drawn up through the packing by convection. The cooled water collects at the bottom of the tower in a reservoir and is returned to the process for reuse, and a make-up of water is added to account for the loss by evaporation. Forced draft cooling towers use fans to pass the cooled air through the packing. Although they have a higher operating cost, they are comparatively smaller and more compact than natural draught cooling towers. In mechanical draft cross-flow cooling towers, the air flows horizontally across the downward flowing water. They therefore have a shorter path for the air to flow and allow a greater flow of air for the same power demand as counter-flow forced draft cooling towers. Compartmental reservoirs are used to capture silt and prevent particles from being circulated in the cooled water.

Problem 3.7: Tank Containment

An open tank containing a liquid freely discharges through a small orifice in the side of the tank. Determine the trajectory of the issuing jet if the depth above the liquid is 3 m. The jet descends 3 m to reach the ground.

Solution

The profile of the freely discharging jet from the side of the tank under the influence of gravity will reach its target at a horizontal distance, x, from the tank while falling a vertical distance, y (Figure 3.6):

$$x = Ut \tag{3.29}$$

and

$$y = \frac{1}{2}gt^2 \tag{3.30}$$

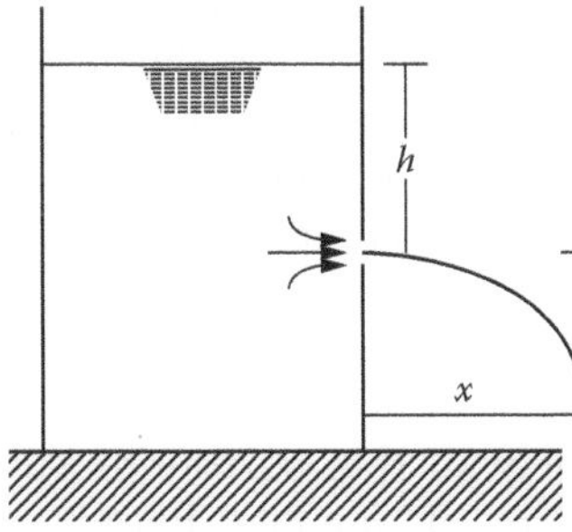

FIGURE 3.6
Discharging Jet

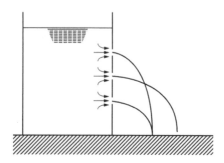

FIGURE 3.7
Discharge from a Tank

Combining Equation 3.29 and Equation 3.30 to eliminate time, t, gives a discharge velocity

$$U = x\sqrt{\frac{g}{2y}} \qquad (3.31)$$

The theoretical discharge velocity is given by conversion of potential energy to kinetic energy:

$$U = \sqrt{2gh} \qquad (3.32)$$

Combining Equation 3.31 and Equation 3.32, the distance travelled by the jet to reach its target is therefore

$$x = 2\sqrt{hy} = 2\times\sqrt{3\times3} = 6 \text{ m} \qquad (3.33)$$

What Should We Look Out For?

The distance travelled from the tank reaches its furthest extent when the orifice is located at half depth. This provides the necessary head for discharge velocity and provides sufficient elevation to permit the jet to extend from the tank. Above or below this compromises either the available head or the elevation for fall as illustrated in Figure 3.7.

What Else Is Interesting?

A bund is a containment wall that surrounds storage tanks which may contain harmful, toxic, or flammable liquids that would otherwise require containment in the event of accidental or deliberate loss of containment. Constructed from concrete or masonry, it is designed to be of sufficient depth to typically contain 110% of the capacity of the tank. Care should be taken with the proximity of the wall to the tank since a loss of containment through an orifice at a half-depth in the tank provides the greatest reach of the issuing jet and may breach the wall if positioned too close, and which has been the cause of bund wall failure in the past.

Problem 3.8: Siphon

A siphon tube with an internal diameter of 0.5 cm is used to transfer water from a tank as shown in Figure 3.8. Neglecting frictional losses, determine the pressure at the top of the tube and the rate of flow.

Solution

Siphoning (or syphoning) is the transfer of liquid from one vessel to another at a lower level by means of a tube or pipe whose highest point is above the surface of the liquid in the upper vessel. It is a useful technique when the original, or upper, liquid has a layer of sediment, which is not to be disturbed. The device is started by placing some of the upper liquid at the top by pressure on the upper surface or suction at the outlet. Once this has

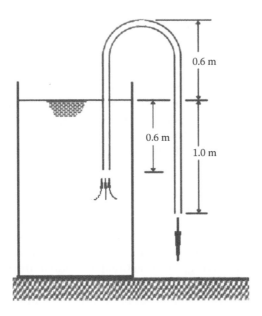

FIGURE 3.8
Operation of a Siphon

been achieved, the liquid will continue to rise to the top point and discharge continuously under gravity to the discharge from the pipe whose outlet is located below the level in the upper vessel. The rate of flow is due to the action of converting the potential energy of the liquid above the outlet into kinetic energy. The mean velocity of the liquid in the tube is therefore

$$U = \sqrt{2gH} = \sqrt{2 \times 9.81 \times 1.0} = 4.43 \text{ ms}^{-1} \qquad (3.34)$$

The rate of flow is

$$\dot{Q} = aU = \frac{\pi d^2}{4} U = \frac{\pi \times 0.005^2}{4} \times 4.43 = 8.7 \times 10^{-5} \text{m}^3\text{s}^{-1} \qquad (3.35)$$

This is the flow rate at the point of starting the siphon. As the operation continues, the capacity of the tank will steadily reduce, and the level will fall. The flow will gradually slow to the point that the level in the tank has fallen to reach the entrance of the siphon tube at which point air will be drawn in and the siphon will stop. Should the outlet be located above the inlet but below the level of the surface in the tank, the siphon will stop when the level reaches the elevation of the outlet.

The pressure at the uppermost point of the tube relative to the surrounding atmospheric pressure will be below atmospheric and is found from

$$p = -\rho g h - \frac{\rho U^2}{2} = -1000 \times 9.81 \times 0.6 - \frac{1000 \times 4.43^2}{2} = -15.7 \text{ kNm}^{-2} \quad (3.36)$$

The negative sign indicates that the pressure is below the atmospheric pressure. A limitation on height of the top point is that the liquid pressure there must not fall below its vapour pressure. Should this happen, bubbles of vapour will come out of the solution, as in boiling, and cause a break in the flow. Note that once the siphon action is under way there is no need for the pressure at the upper surface to be greater than the pressure on the discharging jet.

What Should We Look Out For?

Siphoning is sometimes undesirable where it may be possible for the liquid in a vessel to discharge completely and allow the vessel vapour to escape. To prevent this from occurring, a siphon break is made in the transfer pipe. Air or gas is then deliberately drawn in to break the siphon flow. Vented loops or antisiphon valves are used to prevent back-siphoning and are found in some marine applications as part of an engine exhaust system or plumbing for bilge pumps. They can also be used to prevent water inadvertently returning

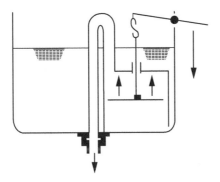

FIGURE 3.9
Operation of a Siphon-Type Cistern

back into the toilet causing an undesirable overflow. While vented loops are simple and very reliable, failure can be catastrophic and result in the potential sinking of ships and boats.

To find the limit for the highest point of the transfer tube, the maximum allowable elevation must correspond to the pressure that exceeds the vapour pressure of the liquid at the operating temperature. Carbonated drinks including beer must therefore be transferred under pressure to ensure that the dissolved carbon dioxide does not come out of the solution.

What Else Is Interesting?

The flushing operation of conventional lavatory cisterns involves siphoning in which the cistern fills with a volume of water sufficient for the necessary flush (Figure 3.9). A float connected to a valve prevents overfilling, and there is an overflow in the event of valve malfunction. The flush is activated by a lever that lifts a volume of water up and over the siphon tube. The flow continues unaided, emptying the cistern until the level of water has reached the entrance to the siphon tube.

Problem 3.9: Water Clock

Determine the geometry of a vessel containing water that discharges freely such that the change in level remains uniform with time.

Solution

The change in level of water in a freely discharging vessel of constant cross section is not linear but proportional to the square root of the depth. An invention attributed to the ancient Egyptians in which the change in level

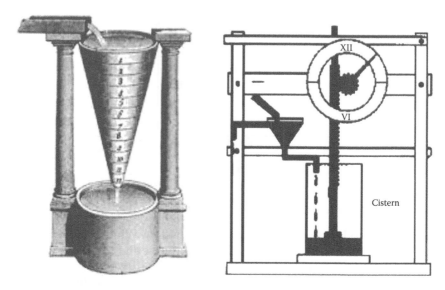

FIGURE 3.10
Examples of Water Clocks (Wikipedia Contributors, "Water Clock," Wikipedia, The Free Encyclopedia, http://en.wikipedia.org/w/index.php?title=Water_clock&oldid=657175335 [accessed May 6, 2015].)

with time is proportional to the depth is the water clock. The clock consists of an open vessel of circular cross section of radius, r, shaped in such a way so as to ensure that the water flows from an orifice in the bottom to cause the depth of water to descend linearly with time. A uniformly scaled ruler serves as a clock "face" (Figure 3.10). The rate of discharge is given in Equation 3.12. If the change in level is constant,

$$\frac{dh}{dt} = -k \tag{3.37}$$

then the ratio of the radius of the vessel is related to the depth of the vessel as

$$r = \frac{C_d a_o (2gh)^{\frac{1}{4}}}{(\pi k)^{\frac{1}{2}}} \tag{3.38}$$

What Should We Look Out For?

Along with sundials, water clocks are the earliest forms of instruments used for measuring time. The Chinese are believed to have been the first to develop such instruments as long ago as 4000 BC. *Water clocks* or *clepsydra* from the Greek *kleptein* and *hydor* meaning "to steal" and "water," have also been in use as long ago as the 16th century BC. They operate using water which is allowed to drip from an upper vessel to the lower vessel in which the level rises, causing a float attached to a stick with notches to turn a gear and with a hand indicates the time. Noted for its accuracy, Galileo used the clepsydra to time his famous experiments involving falling objects. The clepsydra was also once used in courtrooms in Alexandria in which a certain amount of water was measured into a container according to the alleged crime, and then the defendant was permitted to speak until the water ran out.

What Else Is Interesting?

Water has long been used by many early cultures for measuring time and there are plenty of other examples. Time devices have been developed over many centuries with variations of inflow, water wheels, gearing, and escapement mechanisms all with the intention of improving accuracy. The Greeks, Romans, Chinese, Koreans, and Japanese all produced variations and other notable developments being made in Byzantium, Syria, and Mesopotamia. The early forms of clocks were calibrated with the sundial and remained the most accurate form of time measurement until the development of the pendulum clock in the 17th century.

Problem 3.10: Force on a Nozzle

Water flows through a pipe with an internal diameter of 50 mm at a rate of 5 litres per second and discharges through a nozzle that has a diameter of 25 mm. Determine the force on the nozzle required to hold it in place.

Solution

When a fluid flows through a pipe, it carries within it a momentum. This is a quantity given by the mass of a body multiplied by its velocity. Used in the study of dynamics, its quantity is conserved under certain circumstances, and its rate of change gives the amount of force acting on the body.

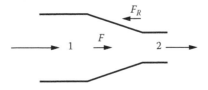

FIGURE 3.11
Force on a Nozzle

As the water discharges, there is a force applied by the water on the nozzle (Figure 3.11). The velocities of the water in the pipe and nozzle are

$$U_1 = \frac{4\dot{Q}}{\pi d_1^2} = \frac{4 \times 0.005}{\pi \times 0.05^2} = 2.54 \text{ ms}^{-1} \tag{3.39}$$

and

$$U_2 = \frac{4\dot{Q}}{\pi d_1^2} = \frac{4 \times 0.005}{\pi \times 0.025^2} = 10.2 \text{ ms}^{-1} \tag{3.40}$$

The total momentum force is

$$F = \rho\dot{Q}(U_2 - U_1) = 1000 \times 0.005 \times (10.2 - 2.54) = 38.3N \tag{3.41}$$

The force required to hold the nozzle in place is found from

$$F = p_1 a_1 - F_R \tag{3.42}$$

The upstream pressure is

$$p_1 = \frac{\rho(U_1^2 - U_2^2)}{2} = \frac{1000 \times (10.2^2 - 2.54^2)}{2} = 48.8 \text{ kNm}^{-2} \tag{3.43}$$

The force on the nozzle to hold it in place is therefore

$$F_R = p_1 a_1 - F = 48,800 \times \frac{\pi \times 0.05^2}{4} - 38.3 = 55.9N \tag{3.44}$$

What Should We Look Out For?

A momentum balance involves the sum of all the forces acting on a moving fluid in one direction and is equal to the difference between the momentum leaving with the fluid per unit time and that brought in per unit time by the fluid. The momentum flow rate of a fluid stream is the product of the mass flow rate and velocity.

What Else Is Interesting?

The momentum of a fluid may be considerable. The force on a pipe bend or elbow, for example, involves resolving the forces in the direction of the fluid approaching and leaving the bend. There are many examples where there has been catastrophic pipe failure as the result of significant momentum transfer effects. Road and rail tankers that transport liquids are also subject to momentum transfer difficulties when braking or negotiating sharp bends. The effect is greater for partly filled tankers but is minimised by using segmented compartments. Such segmented compartments are also used to overcome the motion within offshore oil–water separators on floating, production, storage, and off-loading (FPSO) vessels. FPSOs have storage capacity but are not connected to a sub-sea pipeline. The oil from sub-sea wells is processed and stored aboard the vessel before being off-loaded into shuttle tankers.

Further Problems

1. Explain what is meant by *vena contracta* for a freely discharging orifice.

2. At the end of a fermentation process, a large vat of beer is decanted using a siphon. This consists of a flexible hose with an inside diameter of 25 mm and extends 2 m below the surface of the beer. The vat has a diameter of 3 m, and the beer is transferred to an identical empty vessel below. The discharge of the flexible hose is located 2 m below the point of entry to the hose. Determine the time to transfer 7000 litres of beer. *Answer*: 1742 s

3. If the highest point of the siphon hose is 6 m above the entry to the hose, determine the maximum transfer of beer possible before a vapour block is likely to be encountered. The vapour pressure of the beer may be taken to be 50 kPa. Standard atmospheric conditions (101.3 kPa) may be assumed. The density of beer can be taken to be 995 kgm^{-3}. *Answer*: 127,000 litres

4. A cylindrical storage tank mounted on its axis contains water that flows freely from a horizontal pipe attached at the bottom of the tank. The tank has a diameter of 2 m, and the discharge pipe has a length of 10 m and internal diameter of 5 cm. Determine the time it takes to reduce the water in the tank from an initial level of 4 m to 1 m. The friction factor can be assumed to be 0.005. Neglect all other losses. *Answer*: 1445 s

5. A cylindrical tank is placed with its axis vertical and is provided with a circular orifice 38 mm in the bottom. Water flows into the tank at a uniform rate and is discharged through the orifice. It is found that it takes 107 seconds for the head in the tank to rise from

61 cm to 76 cm and 120 seconds for it to rise from 122 cm to 130 cm. Determine the rate of inflow and the cross-sectional area of the tank assuming that the coefficient of discharge for the orifice is 0.62. *Answer*: 0.0043 m³s⁻¹, 1.26 m²

6. Show that the time to drain a cylindrical tank on its side of radius R and length L through a pipe of radius r extending vertically downward of length l, with laminar flow can be given by

$$t = \frac{-16\mu L l}{\rho g \pi r^4} \int_h^0 \frac{\sqrt{2Rh - h^2}}{h + l}\, dh$$

7. Water is distributed over the packing in a forced convection cooling tower using a channel perforated with 100 holes each with a diameter of 1 cm. If the level of water in the channel is maintained at a depth of 10 cm, determine the rate of water distributed. *Answer*: 6.6 litres per second

8. A cylindrical tank mounted on its axis is used to store a liquid. The tank has a diameter of 4 m and is filled to a depth of 6 m. The tank features a manually operated drain valve at the bottom. If the pressure in the tank is maintained at atmospheric pressure, determine the time to drain the tank through the drain valve if the valve behaves as an orifice with an effective diameter of 25 mm and an assumed coefficient of discharge of 0.6. *Answer*: 47,189 s

9. A straight-sided tank has a surface area of 1 m² and features a hole at the bottom with an area of 10 cm² through which liquid can freely drain. The tank receives a constant flow of water at a rate of 0.02 m³s⁻¹ into the tank. Determine the equilibrium depth. *Answer*: 0.451 m

4

Fluid Friction

Introduction

Most fluids are transported along pipelines that provide effective containment. Fabricated from a range of materials, the pipes themselves are designed to withstand the abrasive or corrosive properties, the pressure and flow. Ducts and other conduits including open channels are also used to transport fluids and, again depending on the properties, are constructed of appropriate materials. The ability to cause a fluid to flow through the pipe or duct is dependent on the applied force. This may be in the form of gravity or the force supplied by a mechanical device such as a pump or compressor. The rate of flow caused is governed by the pressure requirements that are required to be overcome and include elevation, pressure of the fluid, and frictional effects. In determining an appropriate pipe to be installed for a particular fluid system, a detailed analysis of these requirements must be undertaken. While narrow-bore pipes or tubes may be inexpensive compared to large-bore pipes mostly due to the amount of extra materials, a complete analysis should consider operating and capital expenditures. It should be noted that the frictional effects that must be overcome are proportional to the reciprocal of the fifth power of the diameter. In effect, this means that by reducing the diameter by half, the pressure increases by a factor of 32. The economic pipe diameter therefore corresponds to the minimum cost of purchase and operation.

The majority of fluid flow systems operate with turbulent flow. While analytical solutions to flow with laminar or streamline flow were developed over 170 years ago, no single mathematical solution exists for the flow of fluids with turbulent flow. A number of early attempts were made experimentally to correlate data in the form of a friction factor with Reynolds numbers. There are, however, several approaches taken, each of which is widely used, confusing those new to the subject. Each has validity and provides the same final answer. For example, the friction factor developed by the American engineer John Thomas Fanning (1837–1911) is still widely used and is either a fraction or a multiple of other commonly encountered friction factors.

Today, approaches to fluid mechanics involving friction factors as correlations or simple numbers are still widely used. This can provide good insight into the demands of a particular fluid transport system and can be used for design, maintenance, and troubleshooting purposes. More specific and detailed analyses can involve sophisticated computer software packages that permit the analysis of complex pipe network systems or the detailed study of flow through or around complex geometry using computational fluid dynamics.

Another useful approach for analysis involves the use of dimensionless analysis. This is based on forming dimensionless groups from variables that are responsible for influencing an observed dependent variable. Once formed, the groups, without having any measure of scale, are invaluable in predicting behaviour from the small to the large. Dimensionless groups are therefore important in both scale-up and scale-down processes in a wide range of engineering disciplines and, in particular, problems involving the flow of fluids.

Problem 4.1: Connected Reservoir Flow

A supply of water to reservoir C is drawn from two open reservoirs A and B by a pipe network shown in Figure 4.1. The elevation of water in both A and B is 10 m above the level in C. If the valve on the pipe from reservoir B is initially closed, determine the rate of flow from reservoir A. If the valve on pipe BD is then fully opened and assumed to offer no resistance, determine the rate of flow through the three pipes. The density and viscosity of water may be taken to be 1000 kgm^{-3} and 0.001 Nsm^{-2}, respectively. The friction factor may be assumed to be a constant with a value of 0.005.

Pipe	AD	BD	CD
Length (m)	100	200	300
Diameter (mm)	50	80	100

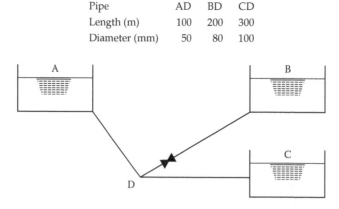

FIGURE 4.1
Pipe System Connecting Three Tanks

Solution

For the first case in which the valve on pipe BD is initially closed, the flow is from reservoir A only. The rate of flow is due entirely to the elevation resisted by the viscous effects along the pipe wall. Since both tanks are open and are assumed to have large capacities such that drainage and changes in levels are not considered, the elevation difference between A and C is therefore the sum of the frictional resistance in the two pipes AD and CD, which expressed as a frictional head loss, is

$$h_{AC} = 4f \frac{L_{AD}}{d_{AD}} \frac{U_{AD}^2}{2g} + 4f \frac{L_{CD}}{d_{CD}} \frac{U_{CD}^2}{2g} \qquad (4.1)$$

where f is the friction factor. Since the volumetric flow rate in both pipes is the same, then

$$\frac{\pi d_{AD}^2}{4} U_{AD} = \frac{\pi d_{CD}^2}{4} U_{CD} \qquad (4.2)$$

Since the diameter of d_{AD} is half that of d_{CD}, then

$$U_{AD} = 4U_{CD} \qquad (4.3)$$

The total head is

$$h_{AC} = 4 \times 0.005 \times \frac{100}{0.05} \times \frac{U_{AD}^2}{2 \times 9.81} + 4 \times 0.005 \times \frac{300}{0.1} \times \frac{(U_{AD}/4)^2}{2 \times 9.81} = 10\,\mathrm{m} \quad (4.4)$$

This gives an average velocity in pipe 1 of 2.12 ms⁻¹. This corresponds to a volumetric flow rate of 0.00416 m³s⁻¹.

For the second case in which the valve on pipe BD is opened, the combined flow is the sum of flow from tanks A and B. That is,

$$\dot{Q}_{CD} = \dot{Q}_{AD} + \dot{Q}_{BD} \qquad (4.5)$$

Applying Bernoulli's equation for both reservoirs between the liquid surfaces, which are effectively stationary, and the junction at D, for which the pressure drops are the same, and assuming that there is no energy loss through the valve or due to entry or exit from the pipes, then

$$4f \frac{L_{AD}}{d_{AD}} \frac{U_{AD}^2}{2g} = 4f \frac{L_{BD}}{d_{BD}} \frac{U_{BD}^2}{2g} \qquad (4.6)$$

The relationship between the velocities in pipes AD and BD is therefore

$$U_{BD} = U_{AD} \left(\frac{L_{AD} d_{BD}}{L_{BD} d_{AD}} \right)^{1/2} = U_{AD} \left(\frac{100 \times 0.08}{200 \times 0.05} \right)^{1/2} = 0.894 U_{AD} \qquad (4.7)$$

Applying Bernoulli's equation to pipe length AC

$$h_{AC} = 4f \frac{L_{AD}}{d_{AD}} \frac{U_{AD}^2}{2g} + 4f \frac{L_{CD}}{d_{CD}} \frac{U_{CD}^2}{2g} \qquad (4.8)$$

where the average velocity in CD is related to the sum of the flow of AD and BD. From Equation 4.5,

$$U_{CD} = \frac{a_{AD} U_{AD} + a_{BD} U_{BD}}{a_{CD}} = \frac{0.05^2 \times U_{AD} + 0.08^2 \times 0.894 \times U_{AD}}{0.3^3} = 0.091 U_{AD} \qquad (4.9)$$

The flow is therefore found from Equation 4.8 to be

$$h_{AC} = 4 \times 0.005 \times \frac{100}{0.05} \times \frac{U_{AD}^2}{2 \times 9.81} + 4 \times 0.005 \times \frac{300}{0.1} \times \frac{(0.091 U_{AD})^2}{2 \times 9.81} = 10\,\text{m} \qquad (4.10)$$

Solving gives a mean velocity through pipe AD of 2.21 ms^{-1}, and therefore a volumetric flow of 4.34×10^{-3} m^{3}s^{-1}.

What Should We Look Out For?

The friction factor is assumed to be constant in each pipe. In reality, however, this is unlikely to be the case. There are many correlations that link friction factor to flow usually by way of the Reynolds number, which can be used to provide a more detailed calculation. The use of charts for friction factor reduces the problem to the use of a trial-and-error iterative process to converge onto a solution.

Problem 4.2: Laminar Flow

A process is to be supplied with glycerol with a specific gravity of 1.26 and viscosity 1.2 Nsm^{-2} from an open storage tank over a distance of 60 m. A pipe with an internal diameter of 12.6 cm is available, and there is a pump that can develop a pressure of 90 kNm^{-2} over a wide range of flows. Determine the maximum glycerol delivery rate that can be achieved by using this equipment if laminar flow is to be maintained.

Solution

This problem is made easier by finding the Reynolds number in terms of mean velocity and deciding the type of flow that exists. In the limit for laminar flow, which corresponds to a maximum Reynolds number of 2,000 for circular pipes, the average velocity in the pipe will not exceed a value of

$$U = \frac{\mu Re}{\rho d} = \frac{1.2 \times 2000}{1260 \times 0.126} = 15.1 \text{ ms}^{-1} \tag{4.11}$$

which corresponds to a maximum flow rate of 0.188 m^3s^{-1}.

The rate of flow of a fluid with laminar flow can be given by the Hagen-Poiseuille equation:

$$\dot{Q} = \frac{\pi}{8\mu} \frac{\Delta p}{L} R^4 \tag{4.12}$$

The maximum flow delivered by the pump therefore corresponds to

$$\dot{Q} = \frac{\pi}{8 \times 1.2} \times \frac{90,000}{60} \times \left(\frac{0.126}{2}\right)^4 = 0.00773 \text{ m}^3\text{s}^{-1} \tag{4.13}$$

which is far less, thereby confirming laminar flow.

What Should We Look Out For?

Glycerol is a viscous Newtonian fluid. Intuition may have suggested laminar flow, but the check with the Reynolds number is useful to confirm this. In fact, in this case, the average velocity in the pipe is 0.62 ms^{-1}, for which the corresponding Reynolds number is 82.

What Else Is Interesting?

The Hagen-Poiseuille equation (Equation 4.12) is a relationship used to determine the rate of flow of a fluid with laminar flow through a horizontal cylindrical tube or pipe. It is named after Gotthilf Heinrich Ludwig Hagen (1797–1884) and Jean Louis Marie Poiseuille (1797–1869), who independently derived this equation in 1839 and 1840, respectively. Hagen was a German physicist who was noted for his contribution to engineering, particularly in the field of hydraulics, and who worked as a civil engineer managing various engineering projects before turning to teaching in Berlin. A French physician and physiologist, Poiseuille was noted for his work on fluids—his major interest was in the flow of blood through the body. By using narrow glass capillaries, he made detailed studies of flow.

Problem 4.3: Tapered Pipe Section

Water flows through a pipe section that tapers from an internal diameter of 200 mm down to 100 mm over a distance of 1 m. Determine the pressure drop over the section for a flow of 0.06 m³s⁻¹. The friction factor may be assumed to be constant with a value of 0.005.

Solution

The pressure gradient due to friction through a pipe is commonly given by the Darcy equation in the form

$$\frac{dp}{dL} = \frac{2f\rho U^2}{d} \tag{4.14}$$

This can be alternatively expressed in terms of flow rate as

$$\frac{dp}{dL} = \frac{32f\rho \dot{Q}^2}{\pi^2 d^5} \tag{4.15}$$

From the geometry of the tapered section, the reduction in the internal diameter is related to length by

$$d = 0.2 - 0.1L \tag{4.16}$$

The total pressure drop over the section is therefore found by integration:

$$\int_0^p dp = \int_0^1 \frac{32f\rho \dot{Q}^2}{\pi^2 (0.2 - 0.1L)^5} dL \tag{4.17}$$

to give

$$\Delta p = \frac{32\,f\rho \dot{Q}^2}{-0.1\pi^2}\left[\frac{(0.2-0.1L)^{-4}}{-4}\right]_0^1 \qquad (4.18)$$

That is,

$$\Delta p = \frac{32\times 0.005\times 1000\times 0.06^2}{0.4\times 3.14^2}\left[(0.2-0.1\times 1)^{-4}-0.2^{-4}\right]=1.4\;kNm^{-2}\quad (4.19)$$

What Should We Look Out For?

The integration here requires a substitution, which is straightforward. The assumption that the friction factor is constant simplifies the problem. The value used here is not unreasonable for general cases unless detailed information is known concerning the surface roughness of the pipe.

What Else Is Interesting?

Tapered sections are commonly encountered in pipelines, for example, to the suction side of pumps, as well as Venturi meters used in the measurement of flow. They are intended to reduce the pressure at the narrowest section or throat and, by virtue of a similar gentle expansion section, to cause a minimum of permanent energy loss. It is more usual to allow for the pressure drop by way of a standard pressure or head loss. In the case of a Venturi flowmeter, the tapered section into and out of the throat is quantified in the form of a coefficient of discharge rather than by determining the actual pressure drop (see Problem 2.1).

Problem 4.4: Ventilation Duct

A building requires a volume of 3 m³s⁻¹ of fresh air. The air is supplied from a conditioning plant a distance of 120 m away for which the permissible pressure drop along the supply duct is 5 kNm⁻². Determine the required pipe diameter if the variation of friction factor is

Reynolds number (Re)	10^4	2×10^4	10^5	10^6	10^7
Friction factor (f)	0.009	0.008	0.007	0.0069	0.0068

The air has a density of 1.2 kgm⁻³ and viscosity of 1.7×10^{-5} Nsm⁻².

Solution

The pressure drop is given by the Darcy equation expressed in terms of volumetric flow rate (Equation 4.15). Rearranged, the duct diameter is therefore

$$d = \sqrt[5]{\frac{32 f \rho L \dot{Q}^2}{\pi^2 \Delta p_f}} = \sqrt[5]{\frac{32 \times f \times 1.2 \times 120 \times 3^2}{\pi^2 \times 5000}} = 0.841\sqrt[5]{f} \qquad (4.20)$$

This is difficult to solve since the friction factor is dependent on the Reynolds number, which itself is dependent on the diameter. An iterative solution is therefore required for which the Reynolds number is

$$Re = \frac{4\rho \dot{Q}}{\mu \pi d} = \frac{4 \times 1.2 \times 3}{1.7 \times 10^{-5} \times \pi \times d} = 2.7 \times 10^5 d^{-1} \qquad (4.21)$$

The solution gives a diameter of 0.31 m.

What Should We Look Out For?

One way to solve this problem is to guess a value for the diameter and calculate the Reynolds number to obtain the friction factor. From this, the diameter can be determined. An adjustment can be made if the guessed and calculated values differ. No more than three iterations should be required.

What Else Is Interesting?

Air conditioning uses ducts that are not always circular in cross section. In practice, the supply line would probably be a rectangular duct. The commonly used Darcy equation can be used in which the diameter can be conveniently replaced by the hydraulic diameter, which is expressed as four times the flow area to the perimeter of the duct.

Problem 4.5: Flow in Noncircular Ducts

A heat exchanger consists of a square duct with inside dimensions of 600 mm by 600 mm and contains four tubes with outside diameters of 210 mm each (Figure 4.2). A liquid used to transfer heat through the tubes has a density of 970 kgm^{-3} and a viscosity of 8×10^{-4} Nsm^{-2}, which flows around the outside of

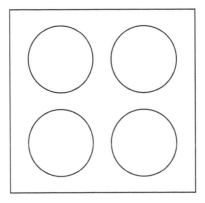

FIGURE 4.2
Duct Cross Section with Four Tubes

the tubes at a rate of 100 litres per second. Determine the pressure drop along the length of the duct by the heat transfer liquid if the absolute roughness of the steel pipes and duct is 0.0046 mm.

Solution

In spite of the difficult geometry, the problem is simplified by noting that the flow within the square section can be equated to the flow in the circular pipe by expressing the flow in terms of the hydraulic diameter of a duct as if it were circular. The hydraulic diameter is often used in calculations of fluids flowing in pipes, enclosed channels, and ducts in which the cross-sectional area is not necessarily circular. It is expressed as the ratio of four times the flow area to the wetted perimeter. In this case, the wetted perimeter is

$$P = 4 \times 0.6 + 4 \times \pi \times 0.21 = 5.04 \text{ m} \tag{4.22}$$

and the flow area is

$$A = 0.6 \times 0.6 - 2 \times \frac{\pi \times 0.21^2}{4} = 0.291 \text{ m}^2 \tag{4.23}$$

The hydraulic diameter is

$$d_H = \frac{4A}{P} = \frac{4 \times 0.291}{5.04} = 0.231 \text{ m} \tag{4.24}$$

The mean velocity of the heat transfer liquid in the duct space is therefore

$$U = \frac{\dot{Q}}{A} = \frac{0.1}{0.291} = 0.344 \text{ ms}^{-1} \tag{4.25}$$

The turbulent flow is confirmed using the Reynolds number:

$$Re = \frac{\rho U d_H}{\mu} = \frac{970 \times 0.344 \times 0.231}{8.0 \times 10^{-4}} = 96,350 \tag{4.26}$$

The relative roughness for the surfaces is

$$\frac{\varepsilon}{d} = \frac{0.0046}{231} = 2 \times 10^{-5} \tag{4.27}$$

This corresponds to a friction factor of 0.0045 for which the pressure drop along the duct is

$$\frac{\Delta p}{L} = \frac{2 f \rho U^2}{d} = \frac{2 \times 0.0045 \times 970 \times 0.344^2}{0.231} = 4.47 \text{ Nm}^{-2}\text{m}^{-1} \tag{4.28}$$

What Should We Look Out For?

This problem is based on the hydraulic diameter. The hydraulic radius could alternatively have been used, which is the cross section or area through which a fluid flows divided by the wetted perimeter:

$$r_H = \frac{A}{P} \tag{4.29}$$

It is often used for open channel flow and for wide channels and represents the depth of the channel. Note that it is not equal to half the equivalent of the hydraulic diameter.

What Else Is Interesting?

The hydraulic diameter is a useful parameter that allows simple calculations to be performed for otherwise complex noncircular geometries. Examples include the flow of fluids through bundles of tubes in shell and tube heat exchangers. These are devices that are used to transfer heat from one medium to another with one heat transfer medium contained within the shell and the other within the tubes. Heat is transferred from one to the other across the tubes. There are many commonly used designs; the simplest is a single-pass-type exchanger in which a cold liquid to be heated flows through the tubes from one side of the exchanger to the other. Steam is used as the heating medium and enters as vapour and leaves as condensate from the bottom. A kettle reboiler is a type of shell and tube heat exchanger in which steam is admitted through the tubes. The choice of hot or cold fluid in the tubes or shell depends upon the application and nature of the fluids, such as their susceptibility to fouling.

Problem 4.6: Valve Test

A valve manufacturer bench tests a new valve to determine the energy loss as a function of the closure. This involves mounting the valve into a plumbing system and measuring the flow rate through the valve and supplying a flow of water with a constant head of 1.5 m from an overhead tank. The valve is fitted to a horizontal length of smooth-walled pipe with a length of 1.5 m and internal diameter of 20 mm. The water has a density of 1000 kgm^{-3} and a viscosity of 0.001 Nsm^{-2}. From the test data in the following table, calculate the loss factor profile defined in head loss form as

$$h_v = k_v \frac{U^2}{2g} \tag{4.30}$$

Valve Closure (%)	Flow (m³min⁻¹)
0	0.084
25	0.060
40	0.036

Solution

This problem involves an iterative procedure in which it is necessary to first determine the friction factor for the pipe. A reasonable value could be assumed to be 0.005 but is worth checking. Using the case in which there is no valve, the head equals the head resistance due to frictional resistance. That is,

$$h = \frac{4fL}{d} \frac{U^2}{2g} \tag{4.31}$$

Expressed in terms of flow rate since the data are already available in this form, the head is

$$h = \frac{32\,fL\dot{Q}^2}{g\pi^2 d^5} \tag{4.32}$$

For smooth-walled pipes, the Blasius equation for the friction factor may apply:

$$f = 0.079\,\mathrm{Re}^{\frac{-1}{4}} \tag{4.33}$$

The Reynolds number is

$$\mathrm{Re} = \frac{\rho U d}{\mu} = \frac{4\rho\dot{Q}}{\pi\mu d} \tag{4.34}$$

For the data given, the guessed values of the flow rate can be used:

Guess	Q (m³s⁻¹)	Re	f	H (m)	Comment
1	0.0015	95,500	0.0045	1.57	Too high
2	0.0014	89,170	0.0046	1.40	Too low
3	0.00146	92,990	0.0045	1.50	Spot on

Note that the friction factor is reasonably constant and the Blasius equation holds true, being within the range of values of 10^4 to 10^5, approximately. Note also that a guessed value of 0.005 would have been incorrect in this case.

For the test valve in place, the head supplied in terms of frictional resistance for the pipe and valve is as follows:

$$h = \frac{4fL}{d}\frac{U^2}{2g} + k_v\frac{U^2}{2g} \tag{4.35}$$

Rearranging, the valve coefficient is

$$k_v = \frac{h - \dfrac{4fL}{d}\dfrac{U^2}{2g}}{\dfrac{U^2}{2g}} \tag{4.36}$$

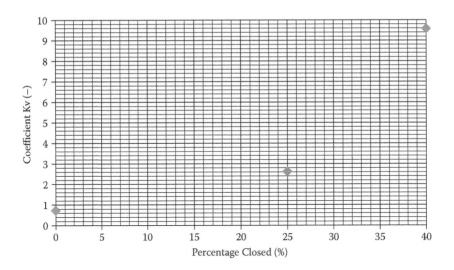

FIGURE 4.3
Valve Test

The values from the data supplied are shown in the following table:

Valve Closure (%)	Q (m³min⁻¹)	k_v
0	0.084	0.71
25	0.060	2.62
40	0.036	9.57

The variation is shown in Figure 4.3.

What Should We Look Out For?

There is a rapid increase in value of the coefficient with the extent to which the valve is closed. In fact, the pressure increases dramatically on valve closure until the point that flow has ceased. The effectiveness of a valve to restrict or control the flow of a fluid has the most significant effect at the point that it is virtually closed.

What Else Is Interesting?

The bench testing of equipment is a commonly used technique to critically evaluate the performance of a device or component prior to installation and operation with the purpose of ensuring that it is fully operable and can perform as expected and with reliability. Bench testing may also be used for equipment that has been withdrawn from service for maintenance and reconditioned with the expectation that it will be returned to service.

Problem 4.7: Flow of a Thick Fluid

A thick oil is to be pumped along a pipeline with an inside diameter of 100 mm at a rate of 7.0 kg⁻¹, determine the pressure drop along the pipeline. If the surface of the wall has a roughness of 0.046 mm, determine the pressure drop. The density of the crude is 890 kgm⁻³ and has a viscosity of 0.002 Nsm⁻².

Solution

The mean velocity for the crude flowing is

$$U = \frac{\dot{m}}{\rho A} = \frac{7}{890 \times \frac{\pi \times 0.1^2}{4}} = 1.0 \, \text{ms}^{-1} \tag{4.37}$$

Assuming that the oil exhibits Newtonian characteristics, the Reynolds number is

$$Re = \frac{\rho U d}{\mu} = \frac{890 \times 1 \times 0.1}{0.002} = 44,500 \tag{4.38}$$

A friction factor can be obtained using either an appropriate chart such as the Moody plot (Figure 4.5) and is found to be 0.000575. Alternatively, an appropriate correlation can be used such as

$$\frac{1}{\sqrt{f}} = -3.6 \log_{10} \left(\frac{6.9}{Re} + \left(\frac{\varepsilon / d}{3.71} \right)^{1.11} \right) \tag{4.39}$$

where the relativeness roughness of the wall is

$$\frac{\varepsilon}{d} = \frac{0.046}{100} = 0.00046 \tag{4.40}$$

Solving gives a friction factor of 0.000564. The pressure drop due to friction is therefore

$$\frac{\Delta p}{L} = \frac{2 f \rho U^2}{d} = \frac{2 \times 0.00564 \times 890 \times 1^2}{0.1} = 100 \, \text{Nm}^{-2}\text{m}^{-1} \tag{4.41}$$

What Should We Look Out For?

There are several widely accepted approaches to determining friction factors. Each has its own validity and provides identical answers. However, it is important to note that friction factor correlations or charts must be used with the appropriate equations. The friction factor developed by American engineer John Fanning (1837–1911) can be related to other friction factors developed by others where

$$f = \frac{\lambda}{4} = \frac{\phi}{4} = \frac{2\tau}{\rho u^2} \tag{4.42}$$

What Else Is Interesting?

Thick fluids do not generally exhibit Newtonian characteristics. Non-Newtonian flow behaviour is presented in more detail in Chapter 9. Briefly, Bingham fluids can be described by

$$\tau = \tau_y + K\dot{\gamma} \tag{4.43}$$

where τ_y is the yield shear stress or yield point and K is the plastic viscosity. Foods that have a high fat content such as mayonnaise, butter, and margarine tend to exhibit Bingham fluid characteristics. This type of fluid is a special case of a Herschel-Bulkley fluid which can be described by

$$\tau = \tau_y + K\dot{\gamma}^n \tag{4.44}$$

Examples include drilling muds.

Problem 4.8: Power Required for Pumping

Crude oil is to be transferred from one tank to another by way of a pump and standard 8-inch Schedule 40 steel pipe at a rate of 4000 litres per minute. The suction line to the pump is 15-m long, and the discharge is a further 180 m (Figure 4.4). The entrance to the feed tank and to the discharge tank are both square-edged, and there is a fully open globe valve in the line. If the crude oil has a specific gravity of 0.88 and dynamic viscosity of 0.085 Nsm^{-2} and the roughness of the steel pipe is 0.046 mm, determine the power

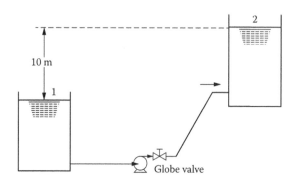

FIGURE 4.4
Tank Transfer

requirement assuming a pump efficiency of 70%. Allow for entrance and exit head losses of 1.5 velocity heads. The valve used to regulate flow has an equivalent length-to-diameter ratio of 340, and the two bends have an equivalent length-to-diameter ratio of 30.

Solution

For the open tanks of a large surface area, the head and hence power requirement required for pumping is dependent only on the elevation and frictional losses. That is,

$$H_P = 10 + h_L \tag{4.45}$$

The total frictional head loss in the system is the sum of the entrance, exit, valve, and pipe friction head losses. This problem specifies the schedule number of the pipe. This is a commonly used non-SI classification for pipes based in terms of their wall thickness. There are 10 schedule numbers in common usage: 10, 20, 30, 40, 60, 80, 100, 120, 140, and 160. In SI units, an 8-inch pipe is 0.2027 m. The average velocity of pumping is therefore

$$U = \frac{4\dot{Q}}{\pi d^2} = \frac{4 \times \dfrac{4}{60}}{\pi \times 0.2027^2} = 2.09 \text{ ms}^{-1} \tag{4.46}$$

The friction factor is found from the relationship with the Reynolds number, which is

$$Re = \frac{\rho U d}{\mu} = \frac{4\rho \dot{Q}}{\mu \pi d} = \frac{4 \times 880 \times \dfrac{4}{60}}{0.0085 \times \pi \times 0.2027} = 43,376 \tag{4.47}$$

The relative roughness of the pipe is

$$\frac{\varepsilon}{d} = \frac{0.000046}{0.2027} = 2.27 \times 10^{-4} \tag{4.48}$$

From the Moody diagram (Figure 4.5), the friction factor is 0.0054.

Allowing for the combined entrance and exit head losses of 1.5 velocity heads and the equivalent length-to-diameter ratio for the valve and two bends gives a head loss due to friction as

$$h_L = 1.5\frac{U^2}{2g} + 2 \times 4f\frac{L_{e(bend)}}{d}\frac{U^2}{2g} + 4f\frac{L_{e(valve)}}{d}\frac{U^2}{2g} + 4f\frac{L_{pipe}}{d}\frac{U^2}{2g} \tag{4.49}$$

That is,

$$h_L = \left(1.5 + 4 \times 0.0054 \times \left(2 \times 30 + 340 + \frac{(15+180)}{0.2027}\right)\right) \times \frac{2.09^2}{2 \times 9.81} = 6.88 \text{ m} \tag{4.50}$$

The globe valve is a device used to regulate the flow of a fluid in a pipe and consists of a flat disc that sits on a fixed ring seat. The disc is movable and allows flow through the valve.

Together with the elevation (Equation 4.45), the head required for pumping is therefore 16.88 m. The power required for pumping is therefore

$$P_o = \frac{\rho g H_p \dot{Q}}{\eta} = \frac{880 \times 9.81 \times 16.88 \times \frac{4}{60}}{0.70} = 13.878 \text{ kW} \tag{4.51}$$

What Should We Look Out For?

It is often sufficient to use a mean value for the friction factor, which is based upon the mean flow rate rather than separate calculations of the friction factor for each section of pipe. If the suction and delivery pipes are of the same diameter, material, and surface finish giving the same friction factor, then the two pipe sections may be treated as one. However, should they be different, then their friction and minor losses will need to be calculated separately.

The surface finish is the degree of roughness of a vessel or pipe surface and is crucial for material durability. A clean surface is essential for providing maximum resistance to corrosion as well as has hygiene and food safety

FIGURE 4.5
Moody Diagram for Friction Factor

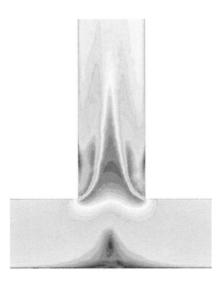

FIGURE 4.6
Velocity Profile of Flow into a T-Piece from Both Directions

implications for food process applications. There are various ways of measuring the surface roughness of pipes and vessels. These include measuring the mean of the absolute values of height difference measured along the surface, the total roughness, the peak-to-valley height in a profile, and the average roughness based on the successive measuring points or an individual peak-and-valley height as seen under a microscope. A profilometer is commonly used to measure a surface profile and uses a diamond stylus, which moves vertically in response to the roughness as it is guided over the surface. Other techniques include visual methods and scanning electron microscopy (SEM).

The equivalent length method is used to determine the permanent energy loss manifested as a pressure drop associated with pipe fittings such as valves, bends, elbows, and T-pieces (Figure 4.6). The equivalent length of a fitting is defined as the length of pipe that would give the same pressure drop as the fitting. Since each size of pipe or fitting requires a different equivalent length for any particular type of fitting, it is usual to express equivalent length as a number of pipe diameters for which this number is independent of pipe. For example, if a valve in a pipe with diameter, d, were said to have an equivalent length of n pipe diameters, then the head loss due to the valve would be the same as that offered by a length of nd of the pipe. Values are determined experimentally.

The entrance and exit losses are irreversible energy losses caused when a fluid enters or leaves an opening such as into or out of a pipe into a vessel. Where there is a sudden enlargement, such as when a pipe enters a larger pipe or vessel, eddies form, and there is a permanent energy loss expressible as a head loss as

$$H_{exit} = \frac{U^2}{2g}\left(1 - \frac{a}{A}\right)^2 \qquad (4.52)$$

For a considerable enlargement the head loss tends to

$$H_{exit} = \frac{U^2}{2g} \qquad (4.53)$$

With a rapid contraction, it has been found experimentally that the permanent head loss can be given by

$$H_{entrance} = k\frac{U^2}{2g} \qquad (4.54)$$

where for a very large contraction, k approaches a value of 0.5.

What Else Is Interesting?

The total head loss along the pipe varies from the inlet to tank 1 to the delivery port of tank 2 (Figure 4.7). There is an immediate head loss into the suction pipe, and head loss continues due to pipe friction and rapidly rises

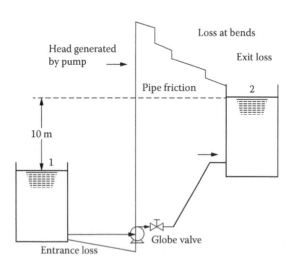

FIGURE 4.7
Total Head in Tank Transfer System

at the pump due to the head supplied to the fluid. It then progressively falls in steps due to the friction along the pipe as well as due to the bends and eventual exit loss into the upper tank.

Problem 4.9: Pipes in Series

A liquid with density of 850 kgm^{-3} and viscosity 0.003 Nsm^{-2} flows through two pipes connected in series at a rate of 200 kg per minute. The two pipes consist of 100 m of horizontal smooth-walled 50 mm bore straight pipe followed by a reducer to 38 mm, with 60 m of horizontal smooth-walled straight pipe. Frictional loss for the reducer amounts to 0.2 velocity heads. Determine the frictional pressure drop across the pipe system.

Solution

The total frictional loss for pipes in series is the sum of all the losses. In this case, the mean velocities of the liquid in the two pipe sections are

$$U_{50mm} = \frac{\dot{m}}{\rho a} = \frac{\dfrac{200}{60}}{850 \times \dfrac{\pi \times 0.05^2}{4}} = 2.0 \text{ ms}^{-1} \tag{4.55}$$

$$U_{38mm} = \frac{\dot{m}}{\rho a} = \frac{\dfrac{200}{60}}{850 \times \dfrac{\pi \times 0.038^2}{4}} = 3.46 \text{ ms}^{-1} \tag{4.56}$$

For the 50 mm pipe section the Reynolds number is

$$Re = \frac{\rho U_{50mm}d}{\mu} = \frac{850 \times 2 \times 0.05}{0.003} = 28,333 \tag{4.57}$$

with a friction factor for smooth-walled pipes of

$$f = 0.079 \, Re^{-1/4} = 0.0062 \tag{4.58}$$

For the 38 mm pipe, the Reynolds number is

$$\text{Re} = \frac{\rho U_{38mm} d}{\mu} = \frac{850 \times 3.46 \times 0.038}{0.003} = 37,252 \tag{4.59}$$

with a friction factor of

$$f = 0.079\,\text{Re}^{-1/4} = 0.0057 \tag{4.60}$$

The total loss of the two pipes including the reducer is therefore

$$h_L = 4 \times 0.0062 \times \frac{100}{0.05} \frac{2^2}{2 \times 9.81} + 0.2 \times \frac{2^2}{2 \times 9.81}$$

$$+4 \times 0.0057 \times \frac{100}{0.038} \frac{2^2}{2 \times 9.81} = 33.82 \text{ m} \tag{4.61}$$

The total pressure drop is therefore

$$\Delta p_f = \rho g h_L = 850 \times 9.81 \times 33.82 = 2.82 \times 10^5 \text{ Nm}^{-2} \tag{4.62}$$

That is 2.82 bar.

What Should We Look Out For?

The head loss for the reducer is based on the higher velocity which is in the 38-mm pipe.

What Else Is Interesting?

This problem is based on resistances in series. In general, this involves a system arranged in a sequence such that the flow through each part is the same and for which the total pressure (or head) across several pipes linked in series is the sum of the pressure drops for each pipe. In other areas of engineering such as in electric circuits in which the circuit elements are arranged in sequence, the same current flows through each, in turn, where the total resistance is equal to the sum of the individual resistances. Likewise, for the flow of heat through a wall composed of several materials, the temperature drop across the entire wall is the sum of the temperature drops for each material.

For systems that involve resistances in parallel, the system is arranged such that a flow is divided through each part such that the resistance is the same in each part. Again, in electric circuits in which resistors are connected, the current divides between them. For pipes in parallel, the flow of fluid divides into each pipe such that the pressure drop resistance of each pipe is the same.

Problem 4.10: Determination of Pipe Diameter for a Given Flow Rate

Water from a reservoir is used to generate electricity at a small hydropower station at an elevation of 500 m below. The water flows freely under the influence of gravity from the reservoir through 5 km of pipe with a 200-mm nominal bore and discharges through a hydro-turbine before discharging into a river. It is proposed to install a second pipe alongside the first to carry an additional flow of water equal to 50%. If the Fanning friction factor in the Darcy equation is assumed to have a constant value of 0.005, determine the minimum inside diameter of the new pipe if the new pipe also flows freely under the influence of gravity. The density of water is assumed to be 1000 kgm⁻³.

Solution

For the existing 200-mm bore pipe which has a head of 500 m, the rate of flow is dependent upon the frictional resistance of the pipe. The Darcy equation (Equation 4.14) in head form is

$$h = \frac{4fL}{d}\frac{U^2}{2g} \tag{4.63}$$

The average velocity in the pipe is therefore

$$U = \sqrt{\frac{2ghd}{4fL}} = \sqrt{\frac{2 \times 9.81 \times 500 \times 0.2}{4 \times 0.005 \times 5000}} = 4.43 \text{ ms}^{-1} \tag{4.64}$$

The rate of flow is

$$\dot{Q} = \frac{\pi d^2}{4}U = \frac{\pi \times 0.2^2}{4} \times 4.43 = 0.1506 \text{ m}^3\text{s}^{-1} \tag{4.65}$$

For the proposed pipe that carries 50% of this flow, the diameter is

$$d = \sqrt[5]{\frac{32 fL\dot{Q}^2}{\pi^2 gh}} = \sqrt[5]{\frac{32 \times 0.005 \times 5000 \times \left(\dfrac{0.1506}{2}\right)^2}{\pi^2 \times 9.81 \times 500}} = 0.098 \text{ m} \tag{4.66}$$

which is an inside diameter of approximately 100 mm.

What Should We Look Out For?

If the friction factor is not known, it will be necessary to determine the flow that uses a trial-and-error approach to obtain the friction factor.

What Else Is Interesting?

The head, or indeed pressure drop, along the pipe is inversely proportional to the fifth power of the inside diameter. This means that a decrease in pipe diameter will increase the pressure by a factor of 32 times. For example, if the pressure along a 100-mm id pipe is 1 bar, the pressure along a pipe of half the diameter with the same flow rate is therefore 32 bar. This is easily shown as follows:

$$p_2 = p_1 \left(\frac{d_1}{d_2} \right)^5 = 1 \times \left(\frac{0.1}{0.05} \right)^5 = 32 \qquad (4.67)$$

This may have significant implications when selecting pipe bores for a particular purpose for which there may be insufficient pumping capability.

Problem 4.11: Drainage through a Horizontal Pipe

Water flows from a cylindrical tank that is 1 m in diameter through a horizontal pipe with a 5 cm internal diameter and is 35 m in length. Determine the time it takes to lower the water surface in the tank from 2.25 m to 0.25 m above the open end of the pipe. Allow for entry and exit losses and assume a friction factor of 0.005.

Solution

The head losses along the pipe including the combined entry and exit losses are

$$h = \frac{4fL}{d} \frac{U^2}{2g} + 1.5 \frac{U^2}{2g} \qquad (4.68)$$

As the tank discharges, the change in capacity for the straight-walled tank is such that only the depth changes with time—that is,

$$aU = -A_t \frac{dh}{dt} \tag{4.69}$$

From Equations 4.68 and 4.69, the time required to reduce the level of liquid can therefore be found by integrating

$$\int_0^t dt = \frac{-A_t}{a\sqrt{\dfrac{2g}{\dfrac{4fL}{d}+1.5}}} \int_{h_1}^{h_2} h\,dh \tag{4.70}$$

to give

$$t = \frac{-2A_t}{a\sqrt{\dfrac{2g}{\dfrac{4fL}{d}+1.5}}}\left(\sqrt{h_1}-\sqrt{h_2}\right)$$

$$\tag{4.71}$$

$$= \frac{-2\times\dfrac{\pi\times 1^2}{4}}{\dfrac{\pi\times 0.05^2}{4}\sqrt{\dfrac{2\times 9.81}{\dfrac{4\times 0.005\times 35}{0.05}+1.5}}}\times\left(\sqrt{0.25}-\sqrt{2.25}\right)=711s$$

What Should We Look Out For?

This problem is simplified by specifying the friction factor. In practice, how-ever, the friction factor is dependent on the rate of flow, which reduces as the tank progressively discharges. A computer iteration process is a simple way to tackle this if a more accurate result is required.

What Else Is Interesting?

As the tank discharges, the pressure drops along the pipe through which the liquid flow decreases. However, if the end of the tank is entirely blocked to stop the flow, the pressure at the end is equal to the static pressure of the liq-uid in the tank. If the pipe were then to be instantaneously "unblocked," there would be a rapid surge in flow and decline to a rate where the forces due to the head in the tank and frictional resistances of flow return to equilibrium.

If the extended pipe was not horizontal but vertical downward, extending from the bottom of the tank, the rate of discharge would be increased. This is because the total head causing flow extends from the top of the liquid down to the discharge point.

Problem 4.12: Shear Stress at a Surface

A Newtonian fluid has a viscosity of 0.01 Nsm^{-2} and flows over a fixed flat surface for which the velocity distribution in the boundary layer is given by

$$\frac{U}{U_o} = 2\left(\frac{y}{\delta}\right) + \left(\frac{y}{\delta}\right)^2 \tag{4.72}$$

where δ is the thickness of the boundary layer, y is the distance from the surface, and U_o is the velocity outside the boundary layer. Determine the magnitude of the shear stress acting on the surface.

Solution

The boundary layer is the region between a surface or wall and a point in a flowing fluid over it where the velocity is at a maximum. Within this region, the movement of the fluid flow is governed by frictional resistance. By convention, the edge of this region is assumed to lie at a point in the flow which has a velocity equal to 99% of the local mainstream velocity. Within the boundary layer, which is laminar inflow, the transfer of heat and mass across it occurs only by molecular diffusion. For the flowing liquid, the shear stress acting on the surface is

$$\tau = \mu\left(\frac{dU}{dy}\right)_{y=0} \tag{4.73}$$

The velocity gradient is therefore

$$\frac{dU}{dy} = \frac{d}{dy}\left(\frac{2yU_o}{\delta} - \frac{y^2U_o}{\delta^2}\right) = \frac{2U_o}{\delta}\left(1 - \frac{y}{\delta}\right) \tag{4.74}$$

At the surface, the shear stress is

$$\tau = \mu\left(\frac{dU}{dy}\right)_{y=0} = \mu\frac{2U}{\delta} = 0.01 \times \frac{2 \times 3}{0.015} = 4 \text{ Nm}^{-2} \tag{4.75}$$

What Should We Look Out For?

It is worth noting that the shear stress at the end of the boundary layer is zero since $y = \delta$ such that

$$\frac{dU}{dy} = 0 \qquad\qquad (4.76)$$

What Else Is Interesting?

The original work is attributed to the French engineer André Lévêque (1896–1930) who noticed that convective heat transfer in a flowing fluid is affected only by the velocity values that are in close proximity to the surface. For flows that have a high Prandtl number, the temperature and mass transition from the surface to bulk temperature takes place across a very thin region close to the surface. The most significant velocities for heat transfer are therefore those inside the boundary layer region.

Problem 4.13: Flow in a Vertical Pipe

A vertical pipe with an internal diameter of 1 cm as shown in Figure 4.8 carries a liquid of density 950 kgm^{-3} and viscosity of 3.25 mNsm^{-2}. Two pressure gauges separated by a distance of 10 m indicate pressures of 200 and 110 kPa, respectively. Determine the direction and rate of flow.

Solution

Pressure gauges at the top and bottom of the vertical pipe indicate the difference in pressure of the liquid contained within the vertical pipe. The gauges

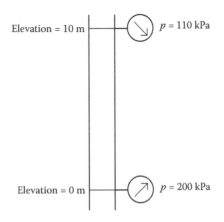

FIGURE 4.8
Flow through a Vertical Pipe

would indicate both static pressure and frictional pressure loss if the liquid were in motion:

$$\Delta p = \rho g h + \Delta p_f \tag{4.77}$$

If the flow is assumed to be laminar on account of the narrow bore, then the Hagen-Poiseuille equation can be given for the pressure drop, expressed in terms of flow rate

$$\Delta p_f = \frac{128\mu L\dot{Q}}{\pi d^4} \tag{4.78}$$

The flow rate is

$$\dot{Q} = \frac{\left(\Delta p - \rho g h\right)\pi d^4}{128\mu L}$$

$$= \frac{\left((200,000 - 110000) - 950 \times 9.81 \times 10\right)\times \pi \times 0.01^4}{128 \times 0.00325 \times 10} \tag{4.79}$$

$$= 2.41 \times 10^{-5}\,\mathrm{m^3 s^{-1}}$$

The two gauges indicate that the flow must be upward since the condition of no flow would therefore exert the difference in static pressure between the two gauges. That is, the pressure difference for no flow is

$$\Delta p = \rho g h = 950 \times 9.81 \times 10 = 93,195\ \mathrm{Nm^{-2}} \tag{4.80}$$

The actual difference is 90,000 Nm-2. However, the difference is very small, and it may be difficult to distinguish this from the needles on the gauges.

What Should We Look Out For?

The flow is assumed to be laminar. A check on the Reynolds number gives

$$\mathrm{Re} = \frac{4\rho\dot{Q}}{\pi\mu d} = \frac{4 \times 950 \times 2.41 \times 10^{-5}}{\pi \times 0.00325 \times 0.01} = 900 \tag{4.81}$$

This confirms laminar flow by virtue of being below a value of 2000 for circular pipes. Had the more general Darcy equation (Equation 4.14) been used, either assuming an approximate friction factor of 0.005 or using an iterative procedure for improved accuracy, would also indicate that the flow is, indeed, laminar.

What Else Is Interesting?

It is often the case that it is not possible to determine the direction of flow of a fluid contained within pipes, or the rate. It is good practice to use labelling where possible.

Problem 4.14: Minimum Pipe Diameter for Maximum Pressure Drop

A horizontal piping system is to be installed to convey water at a flow rate of 250 m³h⁻¹. If the maximum allowable pressure drop per unit length of the pipe is 40 Nm⁻², determine the minimum diameter. The surface roughness of the pipe is 0.14 mm.

Solution

The pressure drop can be given by the general equation

$$\Delta p_f = 2 f \rho U^2 \frac{L}{d} \tag{4.82}$$

where the mean velocity of the water is related to the flow rate by

$$U = \frac{4\dot{Q}}{\pi d^2} \tag{4.83}$$

The diameter is therefore found from

$$d^5 = \frac{32 f \rho L \dot{Q}^2}{\pi^2 \Delta p_f} \tag{4.84}$$

However, this problem is complicated by the friction factor being dependent on the Reynolds number that, in turn, is dependent on the pipe diameter:

$$Re = \frac{4\rho \dot{Q}}{\pi \mu d} = \frac{4 \times 1000 \times 250/3600}{\pi \times 0.0011 \times d} = \frac{80,381}{d} \tag{4.85}$$

The iterative procedure is therefore to assume a value for the friction factor from which an associated pipe diameter can be evaluated from the fifth root calculation. The Reynolds number may then be determined, and using the relative roughness that also depends on the evaluated pipe diameter, a calculated friction factor can then be obtained from the Moody plot. When the calculated and assumed friction factor matches the calculated value, the solution has been reached. In this case, the friction factor of 0.007 satisfies the solution. The corresponding pipe diameter is therefore, in this case, 0.309 m.

What Should We Look Out For?

Where there is a difference in value between the assumed and calculated value, an appropriate adjustment is made to the assumed value. Usually, no more than three values are required that would enable rapid convergence to reach the solution.

What Else Is Interesting?

The problem assumes turbulent flow. Had laminar flow been assumed in which the friction factor is given by $f = 16/Re$, the problem would have quickly led to difficulties confirming that the flow was not laminar flow but turbulent.

Further Problems

1. Water is pumped through a simple piping system that consists of a 1000 m length of pipe with an internal diameter of 150 mm before splitting into two parallel pipes (pipe 1 and pipe 2). Pipe 1 has a length of 500 m and an internal diameter of 100 mm, and pipe 2 has a length of 100 m and an internal diameter of 50 mm. If the rate of water into the piping system is 78.9 m^3h^{-1} and both smaller pipes discharge at the same pressure, determine the flow in the two parallel pipes and the pressure drop. The friction factor may be assumed to be 0.004 in the entry feed pipe and 0.005 in the two smaller pipes. There is no elevation in the piping system. The density of water is 1000 kgm^{-3}, and the viscosity is 0.001 Pa s. *Answer:* 56.6 m^3h^{-1}, 22.3 m^3h^{-1}, 200 kNm^{-2}

2. Water flows freely under the influence of gravity between two open tanks through a pipe with a length of 5 m and an internal diameter of 52 mm and surface roughness of 0.046 mm. Determine the

difference in elevation between the two tanks if the rate of flow is 0.005 m³s⁻¹. *Answer:* 0.5 m

3. Water flows freely under the influence of gravity from an open tank to another open tank at an elevation of 5 m beneath. Determine the rate of flow between the tanks if the pipe is made of coated iron with a surface roughness of 0.012 mm and has an internal diameter of 75 mm. *Answer:* 14.4 m³h⁻¹

4. Oil with a viscosity of 30 mPa s and specific gravity of 0.87 is to be transferred between two vessels by tube. If the pressure drop along the tube is not to exceed 1 kNm⁻² per metre length, determine the internal diameter of the tube if laminar flow is to be ensured.

5. An available pipeline is to be considered for transporting a liquid residue of density 1047 kgm⁻³ and viscosity 0.0157 Pa s. The pipeline is 1500 m in length and has an internal diameter of 200 mm. The pipeline is horizontal and has no elevation issues. From an inspection of pipe records, it is estimated that the inner surface roughness of the pipe is 0.4 mm. If the pressure drop over the pipeline is limited to 265 kPa due to restrictions on pumping delivery pressure, determine the maximum possible transfer rate. *Answer:* 170 m³h⁻¹.

6. Explain what is meant by the term *entrance length* in relation to flow from a reservoir into a pipe. In which type of flow, laminar or turbulent, is the entrance length longer?

7. Starting with a force balance on an element of fluid with laminar flow in a horizontal pipe of circular cross section, show that the fluid velocity for any radius, *r*, can be given by

$$U_X = \frac{1}{4\mu} \frac{\Delta p}{L} \left(R^2 - r^2 \right)$$

where μ is the viscosity, $\Delta p/L$ is the pressure drop per unit length of pipe, and R is the inner radius of the pipe. State any assumptions made.

8. An oil, which exhibits Newtonian behaviour and has a density of 800 kgm⁻³ and viscosity of 80 mNsm⁻², flows through a horizontal pipe with an inner diameter of 2 cm. If the pressure drop along the pipe is 32 kNm⁻² per metre length, determine the rate of flow of oil through the pipe. *Answer:* 0.0015 m³s⁻¹

9. Explain what is meant by the equivalent length and velocity head approaches to estimating the energy losses in pipe fittings. Determine a relationship between the equivalent length and velocity head

methods in terms of Fanning friction factor, f. Explain why inaccuracies occur between the two methods.

10. Define the Reynolds number, and sketch a graph to illustrate the variation of the Fanning friction factor with Reynolds number across a wide range of flow rates and pipe surfaces. Explain why the friction factor in the laminar region appears as a single straight line.

11. Gasoline fuel at 50°C enters a parallel-pipe network that consists of 50 m of 5-cm-diameter steel pipe and the other branch as 100 m of 10-cm-diameter pipe. If the friction factor is assumed to be the same for both pipes, determine the distribution of flow through the two pipes if the feed is 50 litres per second. *Answer*: 10 litres per second and 40 litres per second

12. The pressures at either end of a smooth-walled pipe with an internal diameter of 15 cm and a length of 1200 m carrying an oil of viscosity 3.28 mPa s and specific gravity 0.854 are 850 kNm^{-2} and 335 kNm^{-2}, respectively. Determine the rate of flow if the evaluation of the pipe increases by 15.4 m. *Answer*: 0.0445 m^{3}s^{-1}

13. Water is to be transferred from an open reservoir to another one 60 m below it through a smooth pipe 2400-m long at a rate of 1 m^{3}s^{-1}. Determine the required internal diameter of the pipe. *Answer*: 1.1 m^{3}s^{-1}

14. A tube with a length of 5 m and internal diameter of 20 mm carries fish oil to a process. For reasons of available space, the tube is inclined at an angle of 23.6° and has a pressure gauge located at either end of the tube. Determine the rate of flow along the tube if the flow of oil is 38 litres per minute and the pressure difference is recorded by the two gauges. The oil has a density of 800 kgm^{-3} and viscosity of 0.08 Nsm^{-2}. *Answer*: 0.711 m^{3}h^{-1}; 15.7 kNm^{-2}

15. Water flows through a conical pipe that narrows from 30-cm diameter to 10-cm diameter. The length of the pipe is 3 m. Determine the frictional head loss for a flow of 0.07 m^{3}s^{-1} if the friction factor may be assumed to be 0.005. *Answer*: 0.3 m

16. A process vessel is to be supplied with glycerol of SG 1.26 and viscosity 1200 mPa s from a storage tank 60 m away. A 12.6-cm internal diameter pipe is available, and a pump that will develop 90 kNm^{-2} pressure difference over a wide range of flows. Determine the rate of flow of glycerol. *Hint*: This problem is made easier by finding the Reynolds number in terms of velocity and deciding which type of flow exists. *Answer*: 0.00773 m^{3}s^{-1}

17. Water at a rate of 1 m^{3}s^{-1} is to be transferred from an open reservoir to another 60 m below under the influence of gravity. Determine the internal diameter of the pipe if it is to be smooth-walled pipe with a length of 2400 m. *Answer*: 1.14 m

18. Hot water at a temperature of 80°C with a corresponding density of 1000 kgm^{-3} and viscosity 4×10^{-4} Nsm^{-2} flows at a rate of 50 m^{3}h^{-1} through the annulus between two horizontal concentric tubes. The outer tube has an inner diameter of 150 mm, and the inner tube has an outer diameter of 100 mm. Determine the pressure drop per unit length due to friction if the surface roughness of the tubing is 0.04 mm. *Answer*: 450 Nm^{-2}m^{-1}

5

Pumps

Introduction

The transport or movement of fluids from one place to another presents a number of challenges depending on the physical properties of the fluid, the volume and pressure to be transported, and other environmental requirements. Numerous means of transporting have been devised over the centuries. From ancient times, water has been raised from wells using buckets and other containers including sacks or bags made from animal skin in which the rate of water raised is dependent on the volume of the sack and the frequency of fill and lift which operates using rope haulage or a balanced fulcrum (Figure 5.1). The Archimedes' screw is another ancient mechanical invention devised for transferring water from a low-lying body of water into elevated irrigation ditches. It is attributed to the Greek mathematician and philosopher Archimedes (287–212 BC) on his visit to Egypt. Today, more sophisticated machines have been invented that can transport a wide variety of fluids from gases to highly viscous and non-Newtonian fluids.

The two major types of fluid-transfer machines or pumps are classified as being positive-displacement for bulk handling or metering, and nonpositive displacement pumps, which are also known as rotodynamic pumps (Figure 5.2). Rotodynamic pumps, which include centrifugal and axial pumps, operate by developing a high liquid velocity (kinetic energy) and converting it to pressure. To produce high rates of discharge, such pumps tend to operate at high speeds, although their optimal efficiency is often limited to a narrow range of delivered flows.

Positive displacement pumps operate by drawing liquid into a chamber or cylinder by the action of a piston with the liquid being discharged in the required direction by the use of check valves. This results in a pulsed flow. Positive displacement pumps are, however, capable of delivering significantly higher heads than rotodynamic pumps. Rotary pumps are another form of positive displacement pump also capable of delivering high heads. The fluid is transported between the teeth of rotating and closely meshing gears or rotors and the pump casing or stator. Unlike reciprocating pumps, the flow

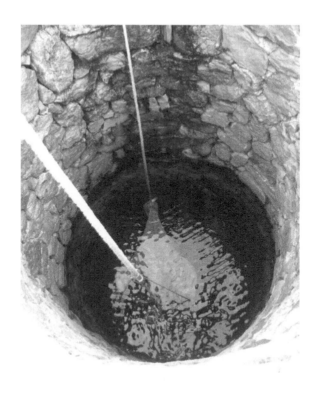

FIGURE 5.1
Lifting Water from a Well (Photo from C.J. Schaschke.)

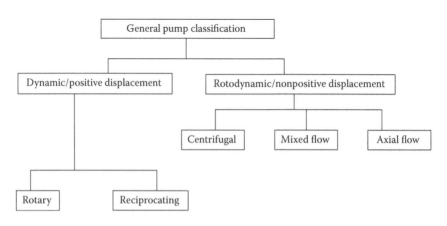

FIGURE 5.2
General Classification of Pumps

is continuous, although rotary pumps tend to operate at lower speeds than rotodynamic pumps, and their physical size tends to be larger.

The decision-making process for the identification and specification of a pump for a particular application follows well-defined steps. To achieve safe, reliable, and efficient operation, the flow delivery and head requirements are the two most obvious factors that need to be considered. The physical characteristics and properties of the fluids in terms of viscosity and lubrication properties, solids content and abrasiveness, as well as corrosion and erosion properties significantly influence the choice of pump type. Also required is a detailed understanding of the required number of pipe or channel networks, number of pumps, and their configuration.

There are many additional calculations and checks that are also required to be made before such an analysis is complete. It is essential, for example, that pumps are correctly located to ensure avoidance of undesirable effects such as the phenomenon of cavitation in centrifugal pumps and separation in reciprocating pumps. Not all pumping installations demand the same level of attention. Pumps required to deliver clean, cold water with modest delivery heads, for example, are more straightforward to design than are pumping systems that involve high temperature and pressure, high viscosity, and abrasive liquids.

In practice, there are numerous additional and important considerations that must also be taken into account. It is important to consider maintenance requirements, implications and likelihood of leaks, availability of spares, economic factors such as capital and operating costs, safety for personnel, and environmental considerations including noise. While the pump specification and selection during the decision-making process may appear complicated, confusing, and possibly conflicting, experience and familiarity with pump selection will eventually result in confidence, thereby reducing the effort required in the successful procurement, installation, and operation of a pump.

Problem 5.1: Pumping of Viscous Liquids

A large quantity of a highly viscous monomer solution with a specific gravity of 0.944 and viscosity of 0.251 Nsm^{-2} is stored in a large tank held under atmospheric pressure. To prevent a known problem of stratification of the solution within the tank over long periods, the monomer is periodically recirculated using a small external pump through a smooth-walled pipeline that is 30 m in length and has an internal diameter of 75 mm. Determine the power requirement of the pump if the recirculation rate is 15 tonnes per hour and the pump has an overall efficiency of 65%. Allow for entrance and exit losses (1.5 velocity heads) in which the pipe has twelve 90° elbows and three open gate valves of 0.7 and 0.15 velocity heads each, respectively.

Solution

In the recirculation process, it is assumed that the liquid is drawn from and returned to the tank such that the head required for pumping is due entirely to overcoming frictional losses. The head required for pumping is therefore

$$H_p = 4f\frac{L}{d}\frac{U^2}{2g} + 1.5\frac{U^2}{2g} + 12 \times 0.7\frac{U^2}{2g} + 3 \times 0.3\frac{U^2}{2g} \tag{5.1}$$

for which the average velocity in the recirculation pipe is

$$U = \frac{\dot{Q}}{a} = \frac{4\dot{m}}{\rho\pi d^2} = \frac{4 \times \dfrac{15,000}{3600}}{944 \times \pi \times 0.075^2} = 1.0 \text{ ms}^{-1} \tag{5.2}$$

The Reynolds number is used to indicate the flow regime:

$$\text{Re} = \frac{\rho U d}{\mu} = \frac{944 \times 1.0 \times 0.075}{0.251} = 282 \tag{5.3}$$

For a circular pipe this is below 2000, which corresponds to laminar flow, for which the friction factor is related to the Reynolds number:

$$f = \frac{16}{\text{Re}} = \frac{16}{0.282} = 0.0567 \tag{5.4}$$

The head for pumping (Equation 5.1) is therefore

$$H_P = \frac{1.0^2}{2 \times 9.81}\left(4 \times 0.0567 \times \frac{30}{0.075} + 1.5 + 8.4 + 0.45\right) = 5.15 \text{ m} \tag{5.5}$$

The actual power required for pumping is

$$P_o = \frac{\dot{m}gH_P}{\eta} = \frac{\dfrac{15,000}{3600} \times 9.81 \times 5.15}{0.65} = 324 \text{ W} \tag{5.6}$$

What Should We Look Out For?

The high viscosity of the liquid may have suggested that the flow would obviously be laminar. It is, however, always worth checking the Reynolds number. The liquid in this case is assumed to be Newtonian. This, however, may not always be the case. The properties of the liquid will have a bearing on the choice of pump particularly if the liquid exhibits non-Newtonian characteristics.

What Else Is Interesting?

While this problem uses the general Darcy equation and the friction factor for laminar flow, the Hagen-Poiseuille equation developed for laminar flow could equally have been used and expressed in head form.

Problem 5.2: Duty Point 1

The pump is required to transfer water from an open tank to a pressurised vessel that operates at a gauge pressure of 75 kNm^{-2}. The transfer line has a length of 20 m with an internal diameter of 0.0525 m and an overall elevation of 10 m. Assuming a friction factor of 0.005, determine the duty point of the pump neglecting all other losses. The performance of a centrifugal pump has the following characteristics:

Discharge ($\times 10^{-4}$) m^{3}s^{-1}	0	33	67	100	133	150	160
Performance head (m)	34.1	33.5	32.0	29.0	22.9	18.3	15.0

Solution

The duty point of a pump depends upon the balance between the system characteristic head and the pump performance head. The performance head, also known as the pump characteristic, provides the relationship between the delivered head and flow. It is used to determine the suitability of the pump to carry out a particular purpose. The duty point is therefore the maximum possible delivery that can be achieved by a particular centrifugal pump to meet the pressure drop (or head) demand of a system. It is represented as the crossover point of the pump characteristic curve with the system characteristic on a pressure-flow-rate curve.

The system characteristic head is composed of both the static head and dynamic head. The static head comprises the potential head as a result of the increase in fluid elevation and the pressure head caused by fluid that may be under pressure. The dynamic head comprises the velocity head as a result of change in fluid velocity. In most systems, the velocity head is negligible. The dynamic head also includes the head loss that is the result of friction and turbulence through the pipes and fittings of the system. The duty point may readily be found by graphical means in which the pump and system head curves are superimposed.

In this problem, the pump is required to overcome elevation, pressure, and frictional losses. Expressed in head form, the head required for pumping is related to the volumetric flow rate as

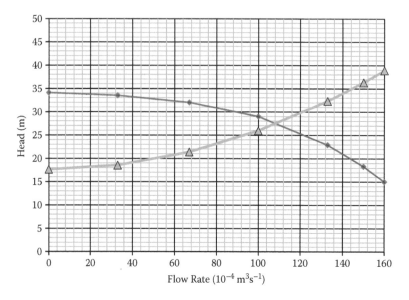

FIGURE 5.3
Duty Point Curve

$$H_p = \frac{p_2 - p_1}{\rho g} + h_2 - h_1 + \frac{4fL}{d}\frac{\left(\dfrac{4\dot{Q}}{\pi d^2}\right)^2}{2g}$$

$$= \frac{75{,}000}{1000 \times 9.81} + 10 + \frac{4 \times 0.005 \times 20}{0.0525}\frac{\left(\dfrac{4 \times \dot{Q}}{\pi \times 0.0525^2}\right)^2}{2 \times 9.81}$$

(5.7)

The system data are presented in Figure 5.3.

The crossover point or duty point flow is found to be 0.0108 m³s⁻¹ with a head of 27.5 m. This is the maximum possible delivery that can be achieved by the pump for this particular system. Having obtained the duty point, it is possible to determine the efficiency, power, and net positive suction head (NPSH) from the pump performance curves.

What Should We Look Out For?

The system characteristic curve in Figure 5.3 is parabolic since the frictional pressure drop through the pipe is proportional to the square of the rate of flow. In this case, there is a head required to be overcome even when there is no flow. This is due to both the elevation causing a static pressure and the pressure exerted by the delivery vessel. In cases of pumping downhill and/or to a receiving tank that operates at a reduced pressure, a curve results that intersects with the y-axis below 0. This indicates the head that needs to be overcome even for no flow. The point where the curve cuts the x-axis indicates the natural flow under the influence of gravity in which the frictional head losses balance the elevation and pressure heads, presuming no nonreturn valves are used.

What Else Is Interesting?

While the duty point signifies the maximum flow and head that can be achieved, higher flows or heads can be achieved by using another pump in parallel (to give a higher flow but the same delivered head) or in series (to give a higher head but the same flow). A particular design of pump to deliver high pressures involves mounting multiple impellers on a single shaft in which the outflow from one impeller feeds into the next (Figure 5.4). In this problem, there are no details regarding valves used to regulate the flow. A valve would, however, provide a pressure (or head) drop that would restrict the flow and alter the position of the duty point.

The performance head provides the relationship between the delivered head and flow. The data presented are for a single pump. A pump selection chart may typically be supplied by manufacturers of pumps featuring several available pumps, and it is used to identify a pump for a particular

FIGURE 5.4
Multiple Impellers on a Single Shaft (Photos from C.J. Schaschke.)

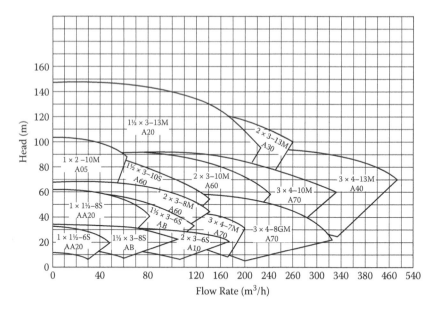

FIGURE 5.5
Pump Selection Chart

duty. The rate of flow is presented on the abscissa and the delivered head or pressure on the ordinate. The performances of a number of pumps are typically presented spanning an acceptable range of efficiencies. Each area corresponds to a name or code that is a combination of information that includes the case number, impeller size, and speed. An example is given in Figure 5.5.

Problem 5.3: Duty Point 2

Determine the duty point, power, efficiency, and NPSH for the three pumps of differing impeller size shown in Figure 5.6 to meet a system characteristic shown below.

Discharge ($\times 10^{-3}$) m³s⁻¹	0	5	10	15	20
System head (m)	16	18	21	25	31

Solution

The chart comprises three pumps with differing impeller diameters. The pump characteristic curve is plotted and superimposed over the three curves as shown in Figure 5.6. The data are summarised in Table 5.1.

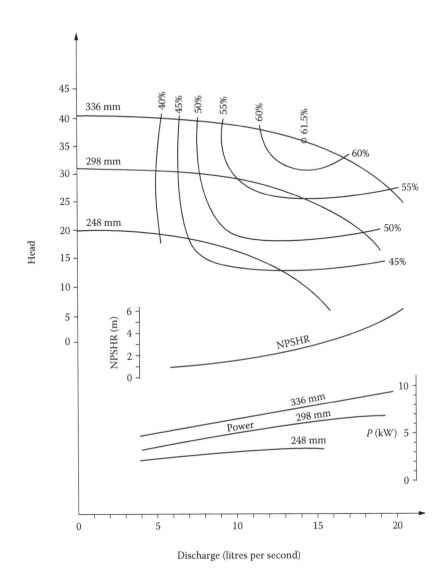

FIGURE 5.6
Pump Performances

TABLE 5.1

Pump Performances

Pump (mm)	Impeller	Flow Rate (ls⁻¹)	Head (m)	Power (kW)	Efficiency (%)	NPSH (m)
1	336	18.9	29	9.3	57	5
2	298	15.1	24	6.5	54	3
3	248	7.8	18	3.0	47	1

What Should We Look Out For?

In general, the most suitable pump has the highest efficiency for the required flow rate. It should be borne in mind that the system curve may change with changes in operating conditions or the passage of time, such as due to the formation of the scale and sedimentation. A pump should therefore be selected whose duty point falls to the right of its peak efficiency point rather than to the left. This will ensure that there is a margin for increase in system head without reduction in efficiency.

What Else Is Interesting?

Where a pump does not run continuously, which is often the case in practice, distinction should be made between the peak duty and the average duty. If the system flow rate fluctuates, the pump may be required to meet the peak flow rate, in which case it will be necessary to choose a pump on this basis. There are, however, many cases where it is only necessary for the pump to meet an average flow rate. In such cases, the pump whose duty point most closely matches the average flow rate requirement will generally be the most economical.

Problem 5.4: Pumping Costs

Determine the cost of running each of the three pumps in Problem 5.3 on a daily basis where the expected quantity of water to be pumped daily is 600 m³. A directly coupled electric motor is to be used with an on/off control, and the electricity cost is 10 cents per kilowatt-hour (kWh). Identify the most suitable impeller for this particular application.

Solution

The total cost for transporting fluids is based on operating costs as well as capital costs. Maintenance costs should also be included which are often independent of the size, type, and number of types, although positive displacement pumps tend to require more attention due to the complexity of the pumps and the tolerances required.

This problem considers only the operating costs. To deliver 600 m³, each pump in Table 5.1 is required to operate for 21.4 hours, 11.04 hours, and 8.82 hours, respectively. For the power demand for each pump, the energy per day is therefore 64.1 kWh, 71.7 kWh, and 82 kWh. At a cost of 10 cents per kWh, this gives a daily cost of $2.88, $3.22, and $3.69, summarised in Table 5.2.

The lowest daily pumping costs correspond to the pump with the 248-mm-diameter impeller.

TABLE 5.2

Cost of Pumping

	Pump 1	Pump 2	Pump 3
Flow rate (ls⁻¹)	18.9	15.1	7.8
Run time (h)	8.82	11.04	21.4
Power input (kW)	9.3	6.5	3.0
Energy (kWh)	82.0	71.7	64.1
Cost ($)	3.69	3.22	2.88

What Should We Look Out For?

In spite of the duty point for pump 3 having the lowest efficiency, this pump provides the lowest running cost. This is because the pump is not required to meet a peak duty but instead to meet the demand of an average delivery of 600 m³ per day. This is equivalent to an average delivery of 6.94 litres per second. Since the duty point of pump 3 is closest to this average duty, it is the best for the average duty requirement. To complete this analysis fully, the capital costs should also be included in this calculation to give the total costs for pumping.

What Else Is Interesting?

A good procedure to ensure that the most appropriate pump is selected is to first determine the desired minimum discharge. The system head at the desired discharge can be calculated in which the system head comprises the frictional losses, pressure head, and static head. The pump selection curves are used before finally determining the actual discharge and head generation.

Problem 5.5: Multi-Phase Pumps and Applications

Identify the pump types that are appropriate for transporting multi-phase mixtures.

Solution

There is an increasing number of multi-phase pumps available that have been developed for transporting liquid products that contain large amounts of gas or vapour. Pumps that handle multi-phase fluids can in a number of cases provide significant equipment and operating savings compared to single-phase flow. In the petroleum industry, for example, various types of pumps have been designed for the transportation of combined oil and gas along with water and sand, where in general, gas and liquid have been separated and transported separately. Multi-phase pumps are therefore able to return economically marginal wells back into production and can handle products ranging from total liquid to total gas. Such pumps are required to be able to switch rapidly between the extremes of total liquid or total gas, or handle slugs of liquid or gas while maintaining full discharge pressure. Additionally, they are required to be capable of handling the heat generated due to gas compression. All require information on the inlet conditions in terms of flow rate of the gas and liquid phases, viscosities, pressures, and temperatures for which most pump manufacturers provide the necessary data required for the pump selection.

Gases are particularly problematic for standard pump designs, which are generally suited to handling solids and liquids. Centrifugal pumps, for example, are commonly used for low-viscosity products and tend to lose efficiency even with low concentrations of entrained gas. Although such pumps can be modified to handle gases, most are generally limited to gas void fractions below 50%.

A wide range of positive-displacement pumps have been successfully used for transporting a wide range of fluids. Rotary screw pumps are ideal for multi-phase applications since they can transport any fluid that can be introduced into the suction passages of their screws and can cope with the heat generation resulting from compressing gases. They find application for liquids that have good lubrication properties although the range of viscosity can vary from a water-like consistency to polymers, and can handle any ratio of gas to liquid. Depending on the materials of construction they can also safely handle a significant amount of solids materials including sand and other abrasive particulates.

Screw-type pumps are another form of positive-displacement pump which are particularly suited to handling difficult liquids as they have intermeshing screws on parallel shafts operating inside close-fitting bores. This design also ensures that the pump's shaft seals are subjected to only the suction pressure, instead of the full discharge pressure, of the pump. The screws are kept in mesh by precision timing gears on each shaft, thereby preventing screw contact. The use of externally lubricated bearings and timing gears allows nonlubricating products to be pumped. Screw pitches can be machined to meet specific requirements permitting direct connection of standard speed drives with reduced power consumption.

Rotary-type positive-displacement pumps can normally move most fluid types that can be introduced into its screw passages. They are generally used for incompressible liquids that flow through the screw passages from the suction to the discharge side as the screws rotate. This action produces trapped volumes of fluid known as *locks*. The number of screw turns on each of the meshing screws determines the number of locks. Close tolerances allow some slip or back-leakage into the suction area. The actual capacity is therefore less than the pump's theoretical displacement. The amount of slip is dependent on the tolerance, number of locks, product viscosity, and differential pressure.

In the transportation of compressible fluids including multi-phase gas–liquid mixtures, with rotary-type positive-displacement pumps the pressure increase occurs toward the last lock nearest to the discharge point. Even for high gas void fractions in excess of 90%, there is usually a sufficient amount of liquid available to provide the necessary sealing in the screw clearances, provided there is a sufficient number of locks and that the differential pressure is not too excessive. The transported liquid will also remove the heat generated by the effects of adiabatic gas compression. For virtually all gas or vapour transportation applications or where there are significant numbers and sizes of gas slugs, special body designs are used to ensure that there is a sufficient amount of liquid available for sealing and cooling.

There are several variations of rotary pumps available that can operate with only a small amount of liquid. For example, screws can be set in a way that allows liquid to be trapped in the bores, or a separating device can be used to separate the liquid and recirculate it back into the pumping chamber. This has the advantage of being able to provide lubrication as well as recirculate liquid flush for the mechanical seals and provide cooling.

Problem 5.6: Centrifugal Pump Scale-Up

A single-stage radial centrifugal pump with an impeller diameter of 131 mm operates at 2500 rpm and produces a head of 12 m and flow rate of 500 litres per minute. It is proposed to replace the pump with a larger geometrically similar pump with an impeller diameter of 150 mm. Determine the impeller speed (rpm) and delivery flow rate for the larger pump if the delivered head is to be only 10 m. Confirm the appropriate type of pump.

Solution

For geometrically similar pumps, dimensionless groups can be used for scale-up (or scale-down) purposes. The dimensionless head, C_H, and capacity coefficients, C_Q, of a centrifugal pump are

$$C_H = \frac{gH}{N^2 D^2} \tag{5.8}$$

and

$$C_Q = \frac{\dot{Q}}{ND^3} \tag{5.9}$$

TABLE 5.3

Centrifugal Pump Scale-Up

Pump Type	Range
Multi-stage centrifugal/positive displacement	$N_S < 0.36$
Single-stage centrifugal (radial)	$0.36 < N_S < 1.10$
Mixed-flow pumps	$1.10 < N_S < 3.60$
Axial-flow pumps	$3.60 < N_S < 5.50$

Using these two coefficients, the flow from the larger size of pump is therefore

$$\dot{Q}_2 = \dot{Q}_1 \frac{N_2}{N_1}\left(\frac{D_2}{D_1}\right)^3 = 500 \times \frac{2000}{2500} \times \left(\frac{0.15}{0.131}\right)^3 = 600 l \, \text{min}^{-1} \tag{5.10}$$

and the impeller speed is

$$N_2 = N_1 \frac{D_1}{D_2}\sqrt{\frac{H_2}{H_1}} = 2500 \times \frac{0.131}{0.15} \times \sqrt{\frac{10}{12}} = 2000 \, \text{rpm} \tag{5.11}$$

The specific speed of the larger pump is therefore calculated from

$$N_S = \frac{N\dot{Q}^{\frac{1}{2}}}{H^{\frac{3}{4}}} = \frac{\frac{2000}{60} \times \sqrt{\frac{0.6}{60}}}{10^{3/4}} = 0.59 \tag{5.12}$$

The specific speed is used to identify the type of pump, as shown in Table 5.3. This corresponds to a single-stage radial centrifugal pump.

What Should We Look Out For?

The specific speed is a classification of centrifugal pump impellers at optimal efficiency with respect to geometric similarity and is useful for the scale-up and selection of centrifugal pumps. It is a measure of pump pressure, head, and speed. Although N_S represents the numerical value for a rotational speed and is usually expressed simply as a number, although it actually has dimensions of $[L]^{3/4}[T]^{-3/2}$. It should therefore be used with caution if non-SI units are used.

What Else Is Interesting?

Unlike the specific speed, the suction-specific speed is a dimensionless number and is used as a measure of centrifugal pump performance for a particular application. Evaluated at the best efficiency point, which corresponds to the maximum efficiency, the suction-specific speed refers to the suction side of the pump and is used to identify issues of cavitation and the type of pump appropriate for a particular application, such as multi- or single-stage, mixed or axial flow. It is calculated from

$$S_n = \frac{N\dot{Q}^{\frac{1}{2}}}{gH^{\frac{3}{4}}}$$

(5.13)

Problem 5.7: Net Positive Suction Head

A hot liquid with a density of 950 kgm^{-3} and vapour pressure of 80 kNm^{-2} is to be transported using a centrifugal pump. The supplied (absolute) pressure of the liquid is 120 kNm^{-2}, and the supplied head is 3 m for which the head loss to the suction side is 0.2 m. The pump manufacturer defines the required net positive suction head to be 4 m. Determine whether cavitation is likely to be a concern.

Solution

Vapour will form in a liquid when the pressure in the liquid is less than the vapour pressure at the liquid temperature. The possibility of this happening is much greater when the liquid is in motion, particularly on the suction side of centrifugal pumps where velocities may be high and the pressure correspondingly low. Having been formed, the vapour bubbles travel with the liquid to regions of higher pressure with eventual collapse, with explosive force giving rise to pressure waves of high intensity. This collapse, or cavitation, occurs on the impeller blades causing noise, vibration, and erosion of the blades, which if not identified and resolved early enough may eventually result in a typically pitted appearance similar to that of corrosion and breakage. Apart from the audible and destructive effect of cavitation, another sign is a rapid decrease in delivered head and pump efficiency. As a remedy, a throttle valve should be placed in the delivery line such that when symptoms of cavitation are identified, the valve should be partially closed, thereby restricting the throughput and thus lowering the velocity.

Cavitation is more likely to occur with high-speed pumps, hot liquids, and liquids with a high volatility—that is, a high vapour pressure. Problems may also be encountered with slurries and liquids with dissolved gas and, in particular, in gas scrubbers that usually operate with liquids saturated with gas. The related phenomenon of separation may occur in reciprocating pumps due to the acceleration of the liquid. This is more likely when the delivery

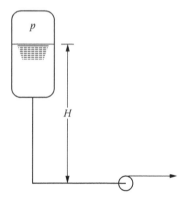

FIGURE 5.7
Net Positive Suction Head

lines are long and the corresponding kinetic energy is large. Apart from providing sufficient NPSH, separation can be avoided by attaching air vessels to
both the delivery and suction lines.

To avoid or overcome problems of cavitation in centrifugal pumps, it is
necessary for the lowest pressure in the pump, which is usually at the eye of
the impeller, to exceed the vapour pressure of the liquid being delivered. The
required net positive suction head (NPSHR) is the positive head necessary to
overcome the pressure drop in the pump and maintain the liquid above its
vapour pressure. This is a function of pump design, impeller size and speed,
and rate of delivery. This value is specified by the pump manufacturer. To
prevent cavitation from occurring, this value must therefore be exceeded by
the available NPSH or NPSHA, which is a function of the system in which
the pump operates. This includes the minimum working gauge pressure
of the vapour on the surface of the liquids, the minimum atmospheric pressure, the vapour pressure of the liquid at the maximum operating temperature, the static head above the pump centre, and the head loss due to friction
and fittings, H_L, as shown in Figure 5.7.

The available NPSH or NPSHA at the pump suction is

$$NPSHA = \frac{p + p_a - p_v}{\rho g} + H - H_L = \frac{120,000 - 80,000}{950 \times 9.81} + 3 - 0.2 = 7.09 \text{ m} \quad (5.14)$$

This value exceeds the required 4 m and so cavitation ought not to be
a problem.

What Should We Look Out For?

The vapour pressure of a liquid is the pressure exerted by the vapour of a
solid or a liquid in which it is in contact and at equilibrium for a specified

temperature. The vapour pressure of pure substances can be obtained from published data or from empirical equations such as the Antoine equation. It should be noted that the vapour pressure of a liquid can be lowered by the presence of a solute, which can both decrease the freezing point and increase the boiling point.

The vapour pressure of liquids rises exponentially with temperature, and while this particular case does not result in cavitation, should the liquid rise in temperature this may result in cavitation. Contributing factors that may unexpectedly lead to cavitation, particularly with volatile or hot liquids, can be attributed to changes in metrological conditions for which atmospheric pressure has been known to fall below 960 mbar. Frictional losses to the pump may also contribute to the likelihood. Suction lines should therefore be kept short and often have a wider diameter than the delivery line. Intercoolers can also reduce the vapour pressure of the liquid.

This particular problem is based on a closed tank that contains a volatile liquid. Open tanks used for the storage of volatile liquids may feature a floating roof. This is a type of cover over the liquid that is allowed to float on the surface, thereby reducing the vapour volume and vaporization and hence material loss.

What Else Is Interesting?

It should be noted that while the available NPSH is a function of the system in which the pump operates, the vapour pressure of the liquid being pumped is not always specifically included in the calculation. In these cases, the vapour pressure head is included such that

$$NPSHA > NPSHR + \frac{p_v}{\rho g} \qquad (5.15)$$

Details of the required NPSH are usually supplied by the pump manufacturer.

Problem 5.8: Centrifugal Pump Scale-Down

The performance of a centrifugal pump is presented below. If another geometrically similar pump with the same size of impeller operates at 70% of the speed of the first, determine the pump characteristics of the new pump.

Flow rate Q ($\times 10^{-3}$ m^{3}s^{-1})	1.47	1.31	1.18	1.00	0.91	0.71	0.48
Head H (m)	1.92	3.67	4.57	6.27	6.48	7.29	8.12

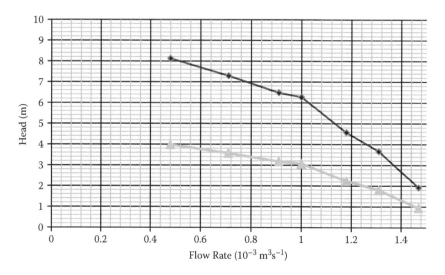

FIGURE 5.8
Pump Scale-Down

Solution

Geometrically similar pumps operate with the same head and capacity coefficients and are related by the head and capacity coefficients, which are dimensionless groups given by Equations 5.8 and 5.9. Combining with $N_2 = 0.7N_1$, therefore, gives $H_2 = 0.49H_1$ and $\dot{Q}_2 = 0.7\dot{Q}_1$. The new pump characteristics are as follows and shown in Figure 5.8:

Flow rate Q ($\times 10^{-3}$ m³s⁻¹)	1.03	0.91	0.83	0.70	0.64	0.50	0.34
Head H (m)	0.94	1.80	2.24	3.07	3.17	3.57	3.98

What Should We Look Out For?

The scale-up or scale-down of equipment is an important part of commercializing a process. Multiple in pumps in series or in parallel can be used to raise the delivery pressure or total flow, respectively, to meet a particular operating demand (Figure 5.9). Dimensionless numbers or groups are a useful way of scaling-up a process since certain heat, mass, and momentum transfer phenomena are independent of scale. The testing of a small-scale process is therefore quick, cost-effective, and reliable such that the experimental information gained can be used on a larger scale for which the equipment is assumed to have a geometric similarity.

What Else Is Interesting?

The head and capacity coefficients are found through a process of dimensional analysis. This involves the Buckingham π theorem, which is a way of

FIGURE 5.9
Centrifugal Test Pumps (Photo from C.J. Schaschke.)

describing how physically meaningful equations that describe observable phenomena involving a number of identifiable variables can be presented as an equation of dimensionless groups in terms of fundamental dimensions such as mass, length, and time. The theorem allows the dimensionless groups to be determined from the variables. The theorem was first proved by the French mathematician Joseph Bertrand (1822–1900) and is named after the American physicist Edgar Buckingham (1867–1940). In this case, the power is a function of the impeller diameter, its rotational speed, fluid flow rate, delivered head (combined with gravitational acceleration), as well as the density and viscosity properties of the fluid.

Problem 5.9: Centrifugal Pump Efficiency

Bench testing of a centrifugal pump operating at 2500 rpm provided experimental data of flow as well as delivery pressure and torque on the rotating shaft. Determine the maximum efficiency of the pump if the force on a torque arm linked to the pump has a length of 17.9 cm as measured from the axis of the rotating shaft. The performance of the pump is as follows:

Flow rate Q ($\times 10^{-3}$ m³s⁻¹)	1.47	1.31	1.18	1.00	0.91	0.71	0.48
Differential press Δp (kNm⁻²)	18.8	36.0	44.8	61.5	63.6	71.5	79.7
Force on arm F (N)	4.6	4.1	4.0	3.7	3.3	3.0	2.4

Solution

A centrifugal pump is a mechanical device used to transport fluids by way of an enclosed impeller rotating at high speed. This widely used device involves the fluid being fed in axially at the centre or eye of the impeller and thrown out in a roughly radial direction by centrifugal action. The large increase in kinetic energy which results is converted into pressure energy at the pump outlet either by using an expanding volute chamber or a diffuser, which can be measured using gauges such as those in Figure 5.10. There are considerable variations in impeller design, but virtually all have blades, which are curved, usually backward to the direction of rotation. This arrangement gives the most stable flow characteristic. The head developed depends not only on the size and rotational speed of the pump but also on the volumetric flow rate.

The bench testing of equipment is routinely used for the validation of new process equipment or for the testing of existing equipment parameters. It can be carried out under controlled conditions, and the data obtained can be analysed in detail. The testing of the performance of centrifugal pumps typically involves mounting the pump such that the flow and pressure across the

FIGURE 5.10
Pressure Gauges on a Centrifugal Pump (Photo from C.J. Schaschke.)

suction and delivery side can be measured. With a wide variation in their design and performance, the most efficient form of operation is limited to a range of flows. The power exerted on the fluid by the pump is the product of the angular velocity of the impeller and its torque:

$$P_o = \omega T \tag{5.16}$$

where the angular velocity is

$$\omega = \frac{2\pi N}{60} \tag{5.17}$$

and the torque is

$$T = Fr \tag{5.18}$$

For a torque arm of 0.179 m, the power input is therefore

$$P_{o(in)} = \frac{2\pi NF \times 0.179}{60} = \frac{NF}{53.3} \tag{5.19}$$

The power delivered by the pump is

$$P_{o(out)} = \dot{Q}\Delta p \tag{5.20}$$

The pump efficiency is therefore

$$\eta = \frac{P_{o(out)}}{P_{o(in)}} \times 100\% \tag{5.21}$$

For the obtained experimental data, the performance of the pump is therefore

Flow rate Q ($\times 10^{-3}$ m^{3}s^{-1})	1.47	1.31	1.18	1.00	0.91	0.71	0.48
Power in $P_{o(in)}$ (W)	215.7	192.3	187.3	173.5	154.8	140.7	112.5
Power in $P_{o(out)}$ (W)	27.6	47.2	52.9	61.5	57.9	50.1	38.3
Efficiency η (%)	12.8	24.5	28.2	35.4	37.4	36.0	34.0

The maximum efficiency is therefore found to correspond to a flow rate of 0.91×10^{-3} m^{3}s^{-1} (Figure 5.11).

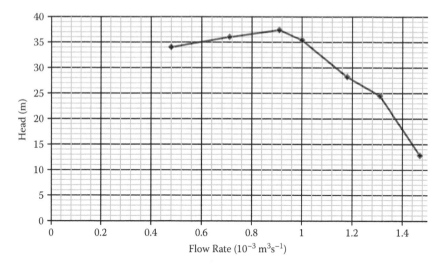

FIGURE 5.11
Pump Efficiency

What Should We Look Out For?

In this particular case, the centrifugal pump has an unusually low efficiency. This may be due to a number of factors, such as blockage or wear of the vanes on the impeller. It may in exceptional circumstances be due to an incorrectly installed impeller, such as having been installed backward. While the pump will function to a certain extent, the performance is often severely impaired.

What Else Is Interesting?

There is a maximum efficiency point for a centrifugal pump known as the *best efficiency point*. This corresponds to the flow rate that will provide the maximum power delivery for the power supplied. If the fluid delivery is throttled, the efficiency of the pump is impaired, and equally at maximum possible rates of flow the delivery pressure is at its lowest. In both extreme cases, the efficiency of the pump is very low, and between is the maximum or best efficiency point. Pump manufacturers usually present their pumps based on the most efficient operation.

Problem 5.10: Reciprocating Pump

A single-acting reciprocating pump has a piston area of 100 cm² and a stroke length of 30 cm and is used to lift water a total elevation of 15 m. If the piston is linked to a crank with a rotational speed of the piston 72 rpm for which the actual quantity of water lifted is 11,000 litres per hour, determine the coefficient of discharge and the theoretical power required to drive the pump.

Solution

A reciprocating pump is a type of positive-displacement device that consists of a piston and cylinder arrangement. The piston is driven by a crankshaft through a connecting rod such that the piston enters and leaves the cylinder expelling and drawing in liquid for transportation in a cyclic fashion. A tight tolerance is required between the piston and cylinder to ensure no leakage. Check valves are used to ensure that the flow is in the desired direction. In their simplest form, these are spheres that lift and seal depending on the flow of the fluid. Multiple cylinders operated out of phase are used to provide a more continuous flow. Variations include the plunger pump and the diaphragm pump.

The volume that is displaced is dependent on the swept volume and the stroke frequency. The flow is pulsed with no flow taking place while the cylinder is recharged.

$$\dot{Q} = aLN = 0.01 \times 0.3 \times 72/60 = 0.0036 \ \mathrm{m^3 s^{-1}} \tag{5.22}$$

The coefficient of discharge is therefore

$$C_d = \frac{11/3600}{0.0036} = 0.85 \tag{5.23}$$

The power is

$$P_o = \rho g \dot{Q} H = 1000 \times 9.81 \times \frac{11}{3600} \times 15 = 450 \ W \tag{5.24}$$

What Should We Look Out For?

The discharge from a reciprocating pump with a single chamber is not steady but pulsed. To provide a smoother flow, multiple chambers can be used, operating out of phase with one another such that there is a continuous flow from a common discharge. To further smooth the flow, an accumulator can be used on the delivery side of the pump. This is a device that can also be used to prevent the destructive effects of water hammer from occurring, which is the violent and potentially damaging shock wave in a pipeline caused by the sudden change in flow rate such as by the rapid closure of a valve. The accumulator consists of a vessel located on the pipe close to the pump with a nonreturn valve preventing return flow back to the pump. The vessel contains a gas or a bladder bag, although some use springs. As the pump discharges, some of the fluid enters the accumulator compressing the gas or spring. At the point of valve closure, the gas or spring expands allowing the accumulated volume to discharge through the pipe. Another way of avoiding water hammer is to control the speed of valve closure, lowering the pressure of the fluid or lowering the fluid flow rate.

What Else Is Interesting?

The human heart is a special form of reciprocating pump in which blood is pumped to the body and the lungs in pulses (heat beats) by the regular and continuous contraction of the heart muscle by flow of electrical stimulation causing expulsion of blood from the ventricles that repeats itself in a cycle. The flow is smoothed by virtue of the elastic properties of the veins and arteries expanding and contracting.

Further Problems

1. A centrifugal pump has an impeller diameter of 200 mm which rotates at 2500 rpm delivering a head of water of 15 m and a flow of 20 litres per second. It is decided to increase the speed of the pump to improve the performance. Determine the new head and flow. *Answer*: 21.6 m and 18 ls^{-1}

2. Describe the types of positive displacement pumps which can be used for liquid transport and their merits.

3. Sketch the rate of flow that discharges from both simplex and duplex reciprocating pumps, and explain the significance of air vessels (accumulators) attached to both suction and delivery lines.

4. Sketch the variation of flow with time for the following reciprocating pump chamber arrangements: single cylinder, duplex, triplex, and quadruplex.

5. Explain what is meant by *cavitation* and the significance of specific speed in centrifugal pump calculations.

6. Construct a flowchart for the decision-making process for the selection of a pump to transfer a process liquid through a pipe.

7. The bench test on the performance of a small centrifugal pump with an impeller diameter of 10 cm at 2500 rpm delivers a flow of water of 20 litres per second with a head of 30 m. Based on this test, determine the delivery flow and head for a geometrically similar pump with an impeller diameter of 15 cm and speed of 3000 rpm. *Answer*: 97.2 m, 0.081 litres s^{-1}

8. A cylindrical tank with a circular cross section with a diameter of 10 m and mounted on its axis tank contains a volatile liquid with a vapour pressure of 20 kNm⁻². The tank, maintained at atmospheric pressure, is to be emptied by way of a temporary centrifugal pump to be located 50 cm above the surface of the liquid when full to a depth of 6 m with a pipe attached to the suction side of the pump which extends down vertically into the tank. If the head losses associated with the pipe amount to 0.5 m and the density of the liquid is 1013 kgm⁻³, determine whether cavitation is likely to be a problem if the required net positive suction head is 4.5 m. *Answer*: NPSHA is 7.2 m.

9. A process receives a liquid hydrocarbon mixture with a density of 800 kgm⁻³ and a viscosity of 1.7×10^{-3} Nsm⁻², at a rate of 5 kgs⁻¹ from a storage vessel that operates at a gauge pressure of 2 bar at a distance of 2 km away. The connecting pipeline available is horizontal and has an internal diameter of 75 mm discharging submerged into a receiving vessel, which is held at a gauge pressure of 5 bar. The level of the liquid in the storage vessel is to be held constant at 5 m above the ground (datum), and the level of the receiving vessel is 10 m above the datum. Determine the power required for pumping if the efficiency of the pump is 65%. *Answer*: 2.5 kW

10. A mixture of volatile liquids is to be separated in a distillation column. The mixture is stored at a gauge pressure of 0.5 bar and is to be pumped to the column at a rate of 34 tonnes per hour through a pipe 100 m in length and with an internal diameter of 10 cm. The mixture has a density of 800 kgm⁻³ and viscosity of 0.6 mPa s. The pipe, with a surface roughness of 0.046 mm, also contains five bends equivalent to 0.7 velocity heads each and four gate valves equivalent to 0.15 velocity heads each. The column, which operates at (standard) atmospheric pressure, has a feed point 25 m above the level of the liquid mixture in the storage tank. If the overall efficiency of the pump and motor is 60%, determine the power requirement of the pump. Allow for pipe entrance and exit losses as well as losses due to friction. *Answer*: 5.3 kW

11. A storage tank containing a liquid of density of 1000 kgm⁻³ and a viscosity of 0.001 Pa s held at a pressure of 1 barg is pumped to a distillation column operating at 2 barg through 50 m of pipe with an internal diameter of 50 mm and surface roughness 0.05 mm at a rate of 20 m³h⁻¹. The pipe contains four open gate valves and 11 bends. Determine the power required for pumping. The column feed point is 10 m higher than the level in the storage tank. The efficiency of the pump is 60%. *Answer*: 3.05 kW

12. A liquid condensate with a density of 600 kgm⁻³ and viscosity 0.9 mPa s from a distillation column is fed at a rate of 5 kgs⁻¹ from the condensate drum held at a gauge pressure of 12 bar to an isomerisation plant for further processing. The pipeline used is considered to be smooth walled with an inside diameter of 7.5 cm and length of 3000 m discharging into a storage tank held at a gauge pressure of 5 bar at the isomerisation plant. If the level of the condensate in the drum is 3 m above the ground (datum) and the discharge of the pipe into the storage tank at the isomerisation plant is 4 m above the datum, determine the power requirement of the pump and motor assembly if their combined efficiency is 65%. Allow for entrance and exit losses and where the pipe contains forty 90° bends of 0.7 velocity heads each and four gate valves (open) of 0.15 velocity heads each. *Answer*: 19.3 kW

13. A storage tank contains a liquid density of 700 kgm⁻³ and a viscosity of 0.0035 Pa s held at a pressure of 1 barg and is pumped to a distillation column operating at 2 barg through 50 m of pipe with an internal diameter of 100 mm and surface roughness 0.046 mm at a rate of 10 m³h⁻¹. The pipe contains four open gate valves and 11 bends. Determine the power required for pumping. The column feed point is 10 m higher than the level in the storage tank. The efficiency of the pump is 60%. Ignore entry and exit losses. *Answer*: 4.23 kW

14. An enclosed storage tank contains a liquid with a density of 1000 kgm⁻³ and a viscosity of 1 mPa s held at a pressure of 1 barg. The tank is pumped at a rate of 20 m³h⁻¹ to a process that operates at 2 barg. The transfer pipe has a length of 50 m with an internal diameter of 50 mm and surface roughness of 0.05 mm. The pipe also contains four open gate valves and 11 bends. Determine the power required for pumping if the process feed point is 10 m higher than the level in the storage tank. The efficiency of the pump is 60%. *Answer*: 3.04 kW

15. A centrifugal pump is used to transport a liquid with a density of 1200 kgm⁻³ and viscosity of 2.4 mNsm⁻² from an open storage tank to a pressure vessel with an increase in elevation of 7 m and at a rate of 14 m³h⁻¹. The gauge pressure in the vessel is 55 kNm⁻². The pipeline consists of 48 m to 52 mm id and 0.21 mm roughness, and has three 90° elbows and two gate valves. The equivalent lengths of the straight pipe for a gate valve and elbow, respectively, are 7 and 40 pipe diameters. The entrance and exit head loss coefficients are 0.5 and 1, respectively. Determine the power requirement for pumping if the pump efficiency is 70%. *Answer*: 1.1 kW

16. A distillation column operating at a gauge pressure of 70 kNm⁻² receives a liquid feed at a rate of 25 m³h⁻¹ from an open storage vessel. The connecting pipework consists of 40 m of 100 mm inside

a diameter pipe with an absolute roughness of 0.046 mm and has a total of 10 bends and three gate valves. The level of liquid feed in the storage tank is steady at 5 m above the ground (datum), and the feed point on the distillation tower is at a height of 9 m above the ground. If the overall efficiency of the pump and motor is 65%, determine the power requirement of the pump. Allow for entrance and exit losses and where the loss for each bend and valve is equivalent to 0.7 and 0.12 velocity heads, respectively. *Answer*: 1.2 kW

17. A solution of concentrated sulphuric acid with a density of 1650 kgm^{-3} and a viscosity of 8.6 Pa s is to be pumped for 800 m along a smooth 5-cm internal diameter pipe at the rate of 10,000 kg per hour, and then raised vertically 12 m by the pump. Determine the power requirement if the pump efficiency is 50%. *Answer*: 1.744 kW

18. An open water storage tank is pumped to a distillation column operating at 2 barg through 50 m of pipe with an internal diameter of 50 mm and surface roughness of 0.05 mm at a rate of 20 m^3h^{-1}. The pipe contains 4 open gate valves and 11 bends. Determine the power required for pumping. The column feed point is 10 m higher than the level in the storage tank. The efficiency of the pump is 60%. The density of the water is 1000 kgm^{-3} and the viscosity is 0.001 Nsm^{-2}. *Answer*: 3 kW

19. Sketch the variation of flow with time for reciprocating pump chambers that consist of a single cylinder, duplex, triplex, and quadruplex arrangements.

20. A storage tank contains a liquid with a density of 1000 kgm^{-3} and a viscosity of 0.001 Pa s held at a pressure of 1 barg and is pumped to a distillation column operating at 2 barg through 50 m of pipe with an internal diameter of 50 mm and surface roughness 0.05 mm at a rate of 20 m^3h^{-1}. The pipe contains 4 open gate valves and 11 bends. Determine the power required for pumping. The column feed point is 10 m higher than the level in the storage tank. The efficiency of the pump is 60%. *Answer*: 3.05 kW

6

Multi-Phase Flow

Introduction

Multi-phase flow concerns the simultaneous flow of two or more states of matter, whereas single-phase flow concerns the flow of one state of matter whether in the form of a gas, vapour, or a liquid. The simplest combination is a two-phase flow that can refer to the flow of a gas and a liquid, a gas and a solid, a liquid and a solid, or two immiscible liquids usually with one phase being dispersed within the other. Multi-phase flow systems are therefore considerably more complex than single-phase flow. To illustrate the complexity, the simultaneous flow of oil and gas from wells in offshore oil and gas installations requires immediate separation for which the flow also requires the separation of water and sand, further complicating matters. Multi-phase flow systems may therefore involve combinations of gases, vapours, liquids, immiscible liquids, as well as solids.

It is important to distinguish between multi-phase flow and multi-component flow. Multi-component flow involves two or more components and is distinct from the state of matter. A two-component flow system, for example, describes flows in which the phases do not consist of the same chemical substance. Steam–water flow, on the other hand, is an example of a two-phase flow, whereas air–water flow is considered to be a two-component flow. Some types of two-component flows such as liquid–liquid flow consist of a single phase but are frequently referred to as two-phase flows in which one phase is identified as continuous and the other as dispersed.

There are many everyday examples of two-phase flow, and many occur in nature such as clouds, rain, and fog along with smoke, smog, quicksand, dust storms, and mud. A fizzy drink poured through the neck of a bottle is a two-phase flow in which the rate of discharge is limited by the rise in velocity of slug-flow bubbles in the neck. The operation of fire extinguishers used to extinguish fires involves multi-phase flow in the form of sprays, jets, foams, and powders. The fires themselves are also usually the result of a reaction between a solid or a liquid fuel with oxygen in the air to produce smoke and water vapour. Internal combustion engines and rockets are designed to combust two-phase dispersions, while refrigeration processes require both evaporation and condensation cycles. Desalination, steel-making, paper manufacturing, and food processes all involve multi-phase flow processes.

The flow equations of multi-phase systems are more complicated or more numerous than those involving single-phase flow. Described by the same laws of physics, the approach required to solve multi-phase flow problems is dependent on the amount of available information. A convenient and common approach used that requires a minimum of analytical effort involves the correlation of experimental data. In spite of their ease of use, it is essential that such correlations are able to be applied with confidence and certainty to situations similar to those that were used to obtain the original data. With often little insight into fundamental flow phenomena, there may be no opportunity to improve the performance or accuracy of prediction. For example, there is considerable complexity in describing even the simplest of situations such as the motion of a bubble of gas rising through a liquid. The action involves the inertia of the gas and the liquid, the viscosity of the gas and the liquid, the density difference and buoyancy, and the surface tension and surface contamination in the form of dirt, dissolved matter, or surface-active agents. Heat and mass transfer to and from the bubble further complicate matters.

Problem 6.1: Open Channel Flow

A rectangular concrete-lined channel is to be built to transport water from a nearby river to a power station as a supply of make-up water. If the channel is to be 10 m wide with a slope of 1 in 5000, determine the depth of the channel if the flow rate is not expected to exceed 650 m^3 per minute. The roughness of the concrete is 0.014 sm$^{-1/3}$.

Solution

While some multi-phase models that have been developed specifically to describe particular flow systems, others have a level of simplicity such that they are not even able to distinguish between the types of flow regime. For example, the flow of gas with a suspension of droplets or stratified flow of a gas over a liquid may be treated conveniently either as a single mass of flowing fluid, or alternatively as individual phases using separate equations to describe each phase with appropriate consideration of the interaction between them.

In its simplest form, open channel flow is a form of multi-phase flow in which liquid flows as a stratified layer above which air, gas, or a vapour exists, and is common in horizontal pipelines such as sewer pipes. The rate of flow through the open channel can be given by the Chézy formula as

$$\dot{Q} = CA\sqrt{mi} \qquad (6.1)$$

where the Chézy coefficient can be given by the Manning formula as:

$$C = \frac{m^{1/6}}{n} \qquad (6.2)$$

The rate of flow is therefore

$$\dot{Q} = \frac{Am^{2/3}}{n}\sqrt{i} \qquad (6.3)$$

The mean hydraulic depth, m, is the ratio of the flow area to the wetted perimeter. Therefore,

$$\frac{650}{60} = 10H\left(\frac{10H}{20+2H}\right)^{2/3}\sqrt{0.0002} \qquad (6.4)$$

Solving by trial and error gives a depth of 1.45 m.

What Should We Look Out For?

In practice, a greater depth would be used and probably a trapezoidal channel would be preferred. This type of design would provide a greater rate of flow and where the walls are not necessarily solid or firm, can reduce the effect of erosion.

Although not always quoted or defined, the surface roughness, n, has SI units of ms$^{-1/3}$. It is important to note the units, especially if Imperial units are used in the calculation which would give other erroneous answers.

What Else Is Interesting?

The trial-and-error approach to solve this problem is a useful problem-solving technique when a variable cannot be easily separated or where there is insufficient knowledge or information within a problem to reach a solution by analytical means. It involves using a reasoned judgment with further adjustments made based on the effects. A guess-and-check approach is another problem-solving technique that involves obtaining a solution by using conjecture to obtain the answer and then checking that it fits the conditions of the problem. It is useful when there is no knowledge or information contained within the problem to reach a solution by alternative means. It is a widely used method and is very useful for solving differential equations in particular.

Problem 6.2: Channel Flow Optimization

An open channel transports water a distance of 5 km from a reservoir to enter a pipe that feeds directly down into a hydropower station. The power station is 650 m below the entry to the channel. Determine the drop in elevation of the channel necessary to maximise the possible power generation. The channel is semi-circular in cross section with a width of 1 m. The surface roughness of the channel is 0.015 sm$^{-1/3}$, and the efficiency of the turbines is 50%.

Solution

The steady flow of water in a channel with air, a gas, or vapour above is a simple form of stratified flow. The rate of flow through the open channel is given by the Chézy formula given in Equation 6.1, for which the Chézy coefficient, C, can be given by the Manning formula or Gauckler-Manning formula in Equation 6.2. The surface roughness of the channel, n, varies for different surfaces and is found by experiment and where the mean hydraulic diameter is the ratio of the flow area to wetted perimeter and given by

$$m = \frac{A}{P} = \frac{\frac{\pi r^2}{2}}{\pi r} = \frac{r}{2} \tag{6.5}$$

The inclination of the channel is

$$i = \frac{h}{L} \tag{6.6}$$

As the flow of water enters the pipe down into the turbines, the theoretical power that can be generated is dependent on both the rate of flow and head. That is,

$$P_o = \rho \dot{Q} g H \tag{6.7}$$

The head available for power generation is related to the total elevation by

$$H_T = H + h \tag{6.8}$$

If a range of drops in elevation for the channel is between 200 m and 250 m, there is a change in power that can be generated by the turbines.

The optimal drop in elevation of the channel can also be found analytically. The power is related to the head as

$$P_O = \frac{\rho \frac{\left(\frac{A}{P}\right)^{\frac{1}{6}} A}{n} \sqrt{\frac{A}{P} \frac{h}{L}} g(H_T - h)}{\eta} \tag{6.9}$$

which is presented more simply as

$$P_O = K\sqrt{h}(H_T - h) \tag{6.10}$$

Differentiating

$$dP_O = \frac{1}{2} KH_T h^{\frac{-1}{2}} dh - \frac{3}{2} Kh^{\frac{1}{2}} dh \tag{6.11}$$

The maximum is found from

$$\frac{dP_O}{dh} = 0 \tag{6.12}$$

to give

$$h = \frac{H_T}{3} = \frac{650}{3} = 217 \text{ m} \tag{6.13}$$

This corresponds to a flow of 13.7 m³s⁻¹ and a power generation of 116 MW. Note that the dimensions and properties of the channel are not relevant in this calculation.

What Should We Look Out For?

Open channel flow usually involves a conduit carrying a liquid with a free surface and is used for transporting large volumes of water at low velocities. The rate of flow is dependent on the slope of the channel, surface roughness, and dimensions. This problem involves a compromise between moving the water to ensure a good flow to the turbine to generate the power by inclining the channel, but balanced by the loss of head into the feed pipe to the turbine. An example of this problem exists in western Uganda where water is drawn from the Rwenzori Mountains and fed to a small hydropower station at the foot of the mountains.

What Else Is Interesting?

The Chézy formula is a semi-empirical formula that relates the rate of discharge of liquid in an open channel to its dimensions, slope, and surface roughness. The formula was devised by the French engineer Antoine Chézy (1718–1798) who was responsible for designing a canal system to supply water to Paris. The Chézy coefficient was developed further by the French engineer Philippe G. Gauckler in 1867 and further still in 1890 by the Irish engineer Robert Manning (1816–1897). The maximum rate of flow is achieved with an open channel with a trapezoidal cross section. For enclosed pipes of circular cross section carrying a flow, the maximum rate of flow can be found to correspond to a depth of 81% of the diameter of the pipe.

Problem 6.3: Stratified Flow

A horizontal pipe with an internal diameter of 200 mm carries a hydrocarbon liquid with a density of 980 kgm^{-3} at a rate of 56 m^3h^{-1}. The pipe is half full and considered to be a stratified two-phase flow with the vapour of the liquid occupying the upper half of the pipe cross section. If a friction factor of 0.005 is assumed for the liquid and the resistance of the vapour–liquid interface is negligible, determine the frictional pressure drop along the pipe.

Solution

This form of two-phase flow consists of a moving layer of liquid below a vapour. The liquid is assumed to move freely along the pipe, and there is no interaction with the vapour above, which is assumed to be stationary. The pressure drop along the pipe wall by the liquid due to friction is obtained from the general equation

$$\frac{\Delta p_f}{L} = \frac{fP\rho U^2}{2A} \tag{6.14}$$

With the liquid occupying only half the cross section, the area of flow is

$$A = \frac{\pi d^2}{8} = \frac{\pi \times 0.2^2}{8} = 0.0157 \ m^2 \tag{6.15}$$

The mean velocity of the liquid is therefore

$$U = \frac{\dot{Q}}{A} = \frac{\frac{56}{3600}}{0.0157} = 0.991 \text{ ms}^{-1} \tag{6.16}$$

The wetted perimeter, which is the inside surface in contact with the liquid, is

$$P = \frac{\pi d}{2} = \frac{\pi \times 0.2}{2} = 0.314 \text{ m} \tag{6.17}$$

The pressure drop in the pipe per metre length (Equation 6.14) is therefore

$$\frac{\Delta p_f}{L} = \frac{0.005 \times 0.314 \times 980 \times 0.991^2}{2 \times 0.0157} = 48.1 \text{ Nm}^{-2}\text{m}^{-1} \tag{6.18}$$

What Should We Look Out For?

The assumption that the flow is stratified should be checked. The Baker plot is a useful way to determine the likely flow regime that may be present for which the mass fluxes of both the vapour and liquid are determined. In this case, the liquid mass flux is just within the regime of the stratified flow, but waves may be formed within the hydrocarbon liquid such that a high mass flux would likely lead to the phenomenon of plug flow.

What Else Is Interesting?

The hydraulic diameter is expressed as the ratio of four times the flow area to the wetted perimeter. In this case,

$$d_e = \frac{4A}{P} = \frac{4 \times 0.0157}{\pi} = 0.2 \text{ m} \tag{6.19}$$

This is useful where the cross-sectional area may not be circular, such that the general force balance equation for flow can be useful directly.

Problem 6.4: Notches and Weirs

Water discharges from a reservoir over a rectangular weir of breadth 10 m and into a sluice below. When the reservoir is at maximum capacity, the rate of discharge over the weir is 5000 m³h⁻¹, and the distance from the surface

FIGURE 6.1
Triangular Notch (Photo by C.J. Schaschke.)

of the reservoir to the sluice is 2 m. Determine the height of the weir crest and the coefficient of discharge if the drop to the sluice is 1.90 cm for a rate of discharge over the weir of 1000 m³h⁻¹.

Solution

A weir is a vertical obstruction across the path of the liquid over which the liquid discharges and is used to retain a capacity of liquids such as in reservoirs, rivers, or streams. The weir crest is the top of the weir over which the liquid flows for which the rate of flow over the top of a rectangular weir is given by

$$\dot{Q} = \frac{2}{3} C_d W \sqrt{2g} H^{2/3} \tag{6.20}$$

If the height of the surface is measured from the sluice below ($H_T = H + H_C$), the height of the crest is

$$H_C = \frac{H_{T1} - H_{T2} \left(\dfrac{\dot{Q}_1}{\dot{Q}_2} \right)^{3/2}}{1 - \left(\dfrac{\dot{Q}_1}{\dot{Q}_2} \right)^{3/2}} = \frac{2 - 1.9 \times \left(\dfrac{5000}{1000} \right)^{3/2}}{1 - \left(\dfrac{5000}{1000} \right)^{3/2}} = 1.89 \text{ m} \tag{6.21}$$

The coefficient of discharge for the weir, which is the ratio of the actual rate of flow to the theoretical rate, is therefore

$$C_d = \frac{3\dot{Q}}{2W\sqrt{2g}(H_T - H_C)^{2/3}} = \frac{3 \times 5000 / 3600}{2 \times 10 \times \sqrt{2 \times 9.81} \times (2 - 1.89)^{2/3}} = 0.2 \quad (6.22)$$

What Should We Look Out For?

Various shapes and forms of notches and weir are most commonly used, the design or choice of which depends on the application. Triangular and trapezoidal notches are often found such as those controlling the discharge from streams and channels. The flow over a trapezoidal notch, which is equivalent in shape to that of a combined triangular notch and rectangular weir is

$$Q = \frac{8}{15} C_d \sqrt{2g} H^{3/2} \left(H \tan\alpha + \frac{5}{4} B \right) \quad (6.23)$$

where α is the angle of the notch. Equation 6.23 can be readily derived from the rate of flow over a triangular notch, which is given by

$$Q = \frac{8}{15} \tan\alpha \, C_d \sqrt{2g} H^{5/2} \quad (6.24)$$

This can be alternatively derived by considering a portion of the flow over a triangular notch, which has the equivalent geometry of a trapezoidal notch.

What Else Is Interesting?

Weirs are commonly used in rivers and streams to retain a body of water that would otherwise flow rapidly, potentially leading to erosion problems.

Problem 6.5: Two-Phase Oil and Gas Flow

A light oil and methane gas flow under pressure along a pipeline with an internal diameter of 200 mm at respective mass rates of 40 kgs^{-1} and 10 kgs^{-1}. The density of the light oil in the pipeline is 800 kgm^{-3} and the methane is 80 kgm^{-3}. Assuming that the gas is incompressible, determine the total mass flow rate, quality of the mixture, individual volumetric flow rates, and superficial velocities and total mass flux.

Solution

Oils are viscid, unctuous, usually inflammable, chemically neutral liquids that are lighter than, and insoluble in, water and classified as nonvolatile. There are many oils that occur in nature. Mineral oils are mixtures of hydrocarbons and have many uses and include fuels, lubricants, soap constituents, and varnishes, and are also used as the feedstock for the production of many other products.

The total mass flow of oil and gas in the pipe is the sum of the individual flows, which is 50 kgs⁻¹. The quality of the mixture is the fraction of the mass of gas in the total mixture:

$$x = \frac{\dot{m}_g}{\dot{m}} = \frac{10}{50} = 0.20 \tag{6.25}$$

The volumetric flow of the oil is the mass flow of the oil divided by its density as

$$\dot{Q}_o = \frac{\dot{m}_o}{\rho_o} = \frac{40}{800} = 0.05 \text{ m}^3\text{s}^{-1} \tag{6.26}$$

The volumetric flow for the gas is

$$\dot{Q}_g = \frac{\dot{m}_g}{\rho_g} = \frac{10}{80} = 0.125 \text{ m}^3\text{s}^{-1} \tag{6.27}$$

The volumetric flux or volumetric flow rate per unit area is also known as the *superficial velocity*. It should be noted that the superficial velocity of the oil is not the same as the velocity at which the oil phase itself moves. The superficial velocity reflects the relative flow rate. The superficial velocity of the gas also reflects the flow rate of the oil in the pipe for which a proportion of the cross section is occupied by the oil and the remainder by the gas. The symbol j is used to represent superficial velocity. Although the flux is a vector quantity, it is used to represent the scale component in the direction. The superficial velocity of the oil is therefore the flow rate of oil per unit cross section:

$$j_o = \frac{4\dot{Q}_o}{\pi d^2} = \frac{4 \times 0.05}{\pi \times 0.2^2} = 1.59 \text{ ms}^{-1} \tag{6.28}$$

and the superficial gas velocity is

$$j_g = \frac{4\dot{Q}_g}{\pi d^2} = \frac{4 \times 0.125}{\pi \times 0.2^2} = 3.98 \text{ ms}^{-1} \tag{6.29}$$

The total volumetric flux is therefore the sum of the two superficial velocities:

$$j = j_o + j_g = 1.59 + 3.98 = 5.57 \text{ ms}^{-1} \tag{6.30}$$

The mass flux, G, is

$$G = \rho j \tag{6.31}$$

for which the total mass flux is therefore

$$G = \rho_o j_o + \rho_g j_g = 800 \times 1.59 + 80 \times 3.98 = 1590.4 \text{ kgm}^{-2}\text{s}^{-1} \tag{6.32}$$

The gas void fraction is the volumetric fraction of gas within the mixture, which in this case is

$$\alpha = \frac{\dot{Q}_g}{\dot{Q}_g + \dot{Q}_o} = \frac{0.125}{0.125 + 0.05} = 0.714 \tag{6.33}$$

What Should We Look Out For?

It is often convenient, particularly in boiling or condensing applications, to quantify a two-phase or two-component flow system in terms of the parameters given in this problem, such as providing a measure of the fraction of the total mass flow across a given area which is composed of each component. One-dimensional flow assumptions permit generalisation for most cases and can be used as a basis for more detailed three-dimensional analysis for more specific problems.

What Else Is Interesting?

The oil and gas industry is primarily associated with the recovery of liquid and gaseous hydrocarbons from underground reservoirs both onshore and offshore. While oil is mainly used as fuel for transportation purposes, it is primarily used as fuel for domestic and industrial purposes, and for converting into other chemicals such as plastic. In terms of transportation, oil is widely transported in ships, whereas gas is transported in underground, sub-sea, or overland pipelines covering large distances. Oil from offshore installations is also brought onshore by sub-sea pipelines.

This problem expresses the rate of flow as kgs^{-1}. However, the term *barrel* is the usual unit used for the quantity of oil. In the United States, one barrel is equal to a volume of 42 (U.S.) gallons and equivalent to 0.159 m^3. One barrel (British) is equal to 36 imperial gallons and equivalent to 0.163 659 m^3. The abbreviation is bbl.

Problem 6.6: Immiscible Liquids

A separation process involves mixing the immiscible liquids kerosene and nitric acid. Determine the mean density of the mixture if the volume fraction of kerosene is 0.3. The density of kerosene is 900 kgm^{-3} and nitric acid, which is dependent on the strength of the acid, is in this problem 1070 kgm^{-3}.

Solution

Immiscibility is the property of fluids to form distinct and separate phases under all relative proportions. For example, oil and water are immiscible as oil floats on water due to its lesser density. Immiscible fluids are often used in liquid–liquid extraction processes in which a dissolved constituent in one phase is transferred to the other by contacting the two phases. This requires a large interfacial surface that can be readily achieved by stirrers, pulsed columns, and in mixer settlers. In this case, the mean density is the contribution from both phases:

$$\rho = \alpha\rho_1 + (1-\alpha)\rho_2 = 0.3 \times 900 + (1-0.3) \times 1070 = 1019 \text{ kgm}^{-3} \quad (6.34)$$

What Should We Look Out For?

In this problem, one phase is dispersed within the other. The mean density is the total mass of the two phases within a given volume. If the two phases are allowed to separate, the mean density within the given volume does not change, albeit the lighter phase (kerosene) will sit upon the heavier phase (nitric acid). Complete phase separation may not be desirable; consequently, a stirrer or some form of recirculation of the mixture is used to ensure the two phases are well dispersed.

What Else Is Interesting?

An example of this liquid–liquid separation process is to be found in the reprocessing of nuclear fuels. The process involves dissolving spent fuel in nitric acid. The uranium and plutonium nitrates are extracted from the solution by a solution of tri-butyl phosphate in kerosene, and the process is based on the ability of nitrates to form chemical complexes with tri-butyl phosphate (TBP). Other fission products are retained in the aqueous phase, which are then concentrated by evaporation. The uranium and plutonium complexes are backwashed from the kerosene phase into the nitric acid solution. There are various forms of this separation with the purex solvent extraction process currently being used to reprocess spent nuclear fuel to separate plutonium, uranium, and fission products.

Problem 6.7: Quality of a Gas

Determine the gas void fraction of a mixture of gas and liquid if the density of the gas is 5 kgm^{-3} and the liquid is 950 kgm^{-3} for which the quality is 0.25.

Solution

The quality of a vapour is expressed as the percentage of saturation of a vapour. A fully saturated vapour therefore has a quality of 100%, while a totally dry vapour has a quality of 0%. As the ratio of the mass of gas to the total mass of the gas–liquid mixture, the quality is therefore

$$x = \frac{m_g}{m} = \frac{\dfrac{m_g}{V}}{\dfrac{m}{V}} = \frac{\rho_g}{\rho} \tag{6.35}$$

The mean density for the two-phase mixture, irrespective of flow regime, is given by Equation 6.34. The gas void fraction, α, is the proportion of the volume that is occupied by the gas phase, which may be in the form of droplets of liquid dispersed in the gas or as bubbles of gas dispersed in the liquid. The gas void fraction is therefore the average occupation of the gas in the two-phase mixture. Experimentally, the gas void fraction can be determined using isolation valves at either ends of a pipe or, if using a glass or transparent pipe or tube, then photography can be used. Within a pipe or tube, the gas void fraction is therefore

$$\alpha = \frac{V_g}{aL} \tag{6.36}$$

Rearranging, the gas void fraction is

$$\alpha = \frac{\rho_L - \dfrac{\rho_g}{x}}{\rho_L - \rho_g} = \frac{950 - \dfrac{5}{0.25}}{950 - 5} = 0.984 \tag{6.37}$$

What Should We Look Out For?

When considering the gas void fraction, a time-average value is used which involves taking measurements over a sufficient period of time. Values may well fluctuate over time and also be of interest such that at any given time the local fluid may be either gas or liquid. The probability of encountering gas at a given point may be determined using local detection probes and is referred to as the *local void fraction*.

What Else Is Interesting?

The gas void fraction is related to the ratio of the superficial velocity of a gas or vapour to liquid in a two-phase flow in a horizontal pipe, known as the *slip ratio*. The simplest approach to estimating the gas void fraction of a flowing gas–liquid mixture is to assume that the flow is homogeneous—that is, both phases flow have the same velocity:

$$S = \frac{U_g}{U_L} \tag{6.38}$$

In terms of the flow rate of the two-phase mixture and gas void fraction

$$S = \frac{\dfrac{Q_g}{A\alpha}}{\dfrac{Q_L}{A(1-\alpha)}} \tag{6.39}$$

Alternatively, the gas void fraction can be expressed in terms of superficial velocity as

$$\alpha = \frac{j_g}{Sj_L + j_g} \tag{6.40}$$

For homogenous flow, the gas void fraction is

$$\alpha = \frac{j_g}{j_L + j_g} \tag{6.41}$$

The slip ratio is not usually equal to 1.0, and care is therefore required when using a model that includes differences in velocity between the two phases. A number of models have been derived, which range from simple one-dimensional models to empirical correlations as well as more complex phenomenological models. These have been developed to overcome the arbitrary nature of empirical correlations. These are based on first identifying the flow pattern or flow regime and then constructing a detailed model for the given flow configuration. Phenomenological modelling has been developed for all the forms of flow regime possible (plug flow, annular flow, bubble flow, stratified flow, slug and churn flow).

Note that the relative velocity is the velocity of one phase with respect to another. The drift flux is the one-dimensional flux of the gas phase in relation to the liquid velocity (assuming a constant void fraction) and is the observed velocity of a phase as it travels in the direction of the fluid.

Problem 6.8: Flow Regimes in Vertical Pipes

A gas–liquid mixture flows up a tube with respective gas and liquid velocities of 3 ms^{-1} and 1.2 ms^{-1}. If the gas void fraction is 0.3, determine the respective superficial velocities of the gas–liquid mixture and identify the flow regime within the pipe. The densities of the gas and liquid are 3 kgm^{-3} and 1000 kgm^{-3}, respectively.

Solution

An example of this problem are the boiler tubes in power stations, which are designed to produce steam from water and are heated directly using high-temperature combustion gases. In the case of vertical boiler tubes, bubbles of steam form on the inside of the walls. These bubbles detach and rise up through the boiling water in the tube. In general, a gas or vapour flowing up through liquids can occur in various forms depending on the extent of the dispersion of one in the other and the physical properties including density, viscosity, and surface tension, and their relative velocities. Using glass tubes, the resulting flow pattern is often evident from visual or photographic observations. However, it is not usually sufficient to define the regime in this way since additional distinguishing criteria are necessary, such as the difference between laminar and turbulent flow or the relative significance of various forces.

It is considerably easier to restrict classifications to the flow patterns such as bubbly, slug, and annular flow in gas–liquid systems and create further subdivisions into distinct regimes within each of these classifications. The transition from one flow regime to another is denoted by hyphenated expressions such as slug-annular and bubbly-slug flows. Synonyms such as *fog* or *mist* are also used. Figure 6.2 shows a sequence of flow regimes that may occur in a vertical boiler tube.

Bubbly flow involves a continuous liquid phase, and the gas phase is dispersed as small bubbles within the liquid continuum. The bubbles travel with a complex motion within the flow, each with a different velocity. The bubbles may be coalescing but are generally of nonuniform size. The wall of the pipe is in continuous contact with the liquid.

Plug flow, which in vertical systems is often referred to as slug flow, is a flow pattern that occurs when the bubble size tends toward that of the

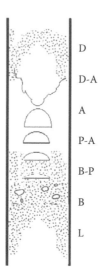

FIGURE 6.2
Flow Regimes in a Vertical Tube (L, liquid; B, bubbly; B-P, bubbly-plug; P-A, plug-annular; A, annular; D-A, drop-annular; D, drop)

channel diameter, and characteristic bullet-shaped bubbles are formed. The gas phase is more pronounced. Although the liquid phase is still continuous, the gas bubbles coalesce and form plugs or slugs that fill most of the pipe cross section. The gas bubble velocity is greater than that of the liquid. The liquid in the film around the bubble may move downward at low velocities. Both the gas and liquid have significant effects on the pressure gradient. The thin film of liquid that surrounds the bubble is known as a *Taylor bubble*. The liquid between Taylor bubbles often contains a dispersion of smaller bubbles.

Churn flow occurs during the transition from a continuous liquid phase to a continuous gas phase when gas bubbles join together and liquid becomes entrained in the bubbles. Although the liquid effects are significant, the gas phase effects are predominant. At higher gas velocities, the Taylor bubbles in the plug flow break down into an unstable pattern in which there is a churning or oscillatory motion of liquid. This type of flow occurs predominantly in wide-bore tubes, while in narrow-bore tubes the region of churn flow is small and may be of less importance. Churn flow with its characteristic oscillations is an important pattern covering a fairly wide range of flow rates.

Annular flow is characterised by liquid travelling as a film on the channel walls with gas flowing through the centre. Some of the liquid may be carried as droplets within the central gas core and is known as *mist flow*. With the pipe wall coated with a liquid film, the gas phase predominantly controls the pressure gradient.

Wispy-annular flow occurs when at very high liquid flow rates, liquid concentrations in the gas core are sufficiently high that droplets coalesce in the gas core, which leads to streaks or wisps to occur instead of droplets. This flow regime corresponds to wispy-annular flow defined in certain flow pattern maps.

The superficial velocity is the velocity of a fluid through a pipe that also contains another fluid of another state and is expressed in terms of the overall cross-sectional area as if no other fluid were present. For the liquid and gas phases, the superficial velocities are therefore

$$j_L = (1-\alpha)U_L = (1-0.3)\times 1.2 = 0.84\,\mathrm{ms^{-1}}$$

$$j_g = \alpha U_g = 0.3\times 3 = 0.9\,\mathrm{ms^{-1}}$$

(6.42)

Figure 6.3 presents the flow regimes based on superficial velocities and experimental data. In this case, the parameters are calculated to be

$$\rho_L\, j_L^2 = 1000\times 0.84^2 = 706\ \mathrm{kgm^{-1}s^2}$$

$$\rho_g\, j_g^2 = 3\times 0.9^2 = 2.43\ \mathrm{kgm^{-1}s^2}$$

(6.43)

From Figure 6.3, this corresponds to the plug flow.

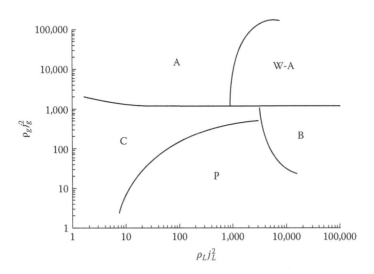

FIGURE 6.3
Flow Regime Map for Two-Phase Flow in a Vertical Pipe (A, annular; W-A, wispy-annular; C, churn; P, plug; B, bubbly)

What Should We Look Out For?

The coordinates of the flow map can alternatively be expressed in terms of mass fluxes of the two phases as

$$\rho_L j_L^2 = \frac{G_L^2}{\rho_L}$$

$$\rho_g j_g^2 = \frac{G_g^2}{\rho_g}$$

(6.44)

Note that the superficial velocity is the volumetric flow of each phase across the entire cross section, and the mass flux is the mass flow per unit flow area.

What Else Is Interesting?

In this problem, it is possible to determine the transition from one phase to another. For example, by raising the gas velocity, the transition to annular flow or wispy-annular flow occurs at $\rho_g j_g^2$ of 100 kgm^{-1}s^2. Presuming the liquid velocity and gas void fraction remains unchanged, this corresponds to a superficial velocity of

$$j_g = \sqrt{\frac{100}{\rho_g}} = \sqrt{\frac{100}{3}} = 5.77\,\text{ms}^{-1}$$

(6.45)

and an actual velocity of

$$U_g = \frac{j_g}{\alpha} = \frac{5.77}{0.3} = 19.2\,\text{ms}^{-1}$$

(6.46)

This velocity is typically encountered with steam in boiler tubes.

Problem 6.9: Vertical Two-Phase Flow

A mixture of air and water flows up a vertical tube with respective mass flows of 0.018 kgs^{-1} and 0.50 kgs^{-1} for which the gas void fraction is measured and found to be 0.88. If the tube has an inside diameter of 25.4 mm, determine the interfacial friction factor between the air and the water.

Solution

The flow regime can be confirmed from the mass fluxes for both the water and air in which

$$G_g = \frac{\dot{m}_g}{A} = \frac{0.018}{\dfrac{\pi \times 0.0254^2}{4}} = 35.5 \ \text{kgm}^{-2}\text{s}^{-1}$$

$$G_L = \frac{\dot{m}_L}{A} = \frac{0.5}{\dfrac{\pi \times 0.0254^2}{4}} = 987 \ \text{kgm}^{-2}\text{s}^{-1}$$

(6.47)

The parameters for the flow regime map are therefore

$$\frac{G_g^2}{\rho_g} = \frac{35.5^2}{1.2} = 1052 \ \text{kgm}^{-1}\text{s}^{-1}$$

$$\frac{G_L^2}{\rho_L} = \frac{987^2}{1000} = 974 \ \text{kgm}^{-1}\text{s}^{-1}$$

(6.48)

From Figure 6.3, this confirms annular flow, which is a two-phase flow regime of a gas and a liquid found in vertical pipes or tubes characterised by a continuous gas core with a wall film of liquid. This flow regime occurs at high gas velocities in which there is often a simultaneous flow of the liquid phase entrained in the gas as a fine dispersion of droplets. In horizontal pipes, the effect of gravity causes the film to become thicker along the bottom of the pipe. As the gas velocity is increased, the film becomes more uniform around the circumference.

The interfacial frictional factor is an important parameter in fluid flow and is used to characterise pressure drop. For two-phase flow, the friction factor is usually defined as the ratio of the interfacial shear stress to the kinetic energy of the vapour phase. A considerable number of relationships have been developed based on experimental data concerning the interfacial friction factor for rough or wavy-annular flow between the liquid and gas. A useful correlation by Wallis (1969) is based on experimental data using a friction factor of the following form:

$$f_i = 0.005\left(1 + 300\frac{\delta}{d}\right)$$

(6.49)

From the geometry of the pipe and annular flow of liquid at the wall with a film thickness of δ, the gas void fraction is therefore

$$\alpha = \frac{A_g}{A} = \frac{\dfrac{\pi(d-2\delta)^2}{4}}{\dfrac{\pi d^2}{4}} = \frac{(1-2\delta)^2}{d^2}$$

(6.50)

Where the film thickness is very small, the ratio of the film thickness to the inside diameter approximates to

$$\frac{\delta}{d} = \frac{1-\alpha}{4} \tag{6.51}$$

The interfacial frictional factor from Equations 6.49 and 6.51 is therefore

$$f_i = 0.005\left(1+75(1-\alpha)\right) \tag{6.52}$$

This gives a value of

$$f_i = 0.005\left(1+75(1-0.88)\right) = 0.05 \tag{6.53}$$

What Should We Look Out For?

The interfacial shear stress between liquids and gases is highly complex such that no single theory exists. This is due to the liquid film, which effectively behaves as a roughened surface, thereby increasing drag. The friction factor therefore depends on the ratio of the mean height of the film to the pipe diameter. For cases where the film is too thin for disturbance waves to be present, the assumption of the liquid film being a roughened surface does not hold.

Assuming that the interfacial shear stress depends on the difference between the gas velocity and a characteristic interface velocity, then for a gas velocity which is considerably greater than the liquid velocity, the liquid velocity may be neglected such that

$$\tau_i = f_i \frac{\rho_g v_g^2}{2} = f_i \frac{\rho_g j_g^2}{2\alpha^2} \tag{6.54}$$

The pressure drop along the pipe then becomes

$$\frac{dp}{dz} = \frac{4\tau_i}{d\sqrt{\alpha}} = \frac{2 f \rho_g j_g^2}{d\alpha^{5/2}} \tag{6.55}$$

If the same gas flow were to fill the pipe, the wall shear stress would be given in terms of the friction factor, f_g, by

$$\tau_{wg} = f_g \frac{\rho_g j_g^2}{2} \tag{6.56}$$

for which the pressure drop would be

$$\frac{dp}{dz} = \frac{4\tau_{wg}}{d} = \frac{2f_g \rho_g j_g^2}{d} \tag{6.57}$$

Combining these equations with the definition of φ_g^2, then

$$f_i = f_g \alpha^{5/2} \varphi_g^2 \tag{6.58}$$

which effectively provides a physical interpretation for the Martinelli parameter φ_g^2.

What Else Is Interesting?

The rough pipe correlations of Nikuradse (1933) and Moody (1944) can also be approximated by the following equation:

$$f_i \approx 0.005\left(1 + 75\frac{\varepsilon}{d}\right) \tag{6.59}$$

This equation is valid over the range of relative roughnesses, ε/d, of 0.001 to 0.03, where ε is the size of a grain of sand used by Nikuradse in his experiments of artificially roughened surfaces in pipes. Wavy-annular film therefore corresponds to an equivalent sand roughness of four times the film thickness.

Problem 6.10: Two-Phase Flow in a Vertical Pipe

A mixture of a gas and a liquid flow together upward through a vertical tube with a length of 20 m and an internal diameter of 30 mm with the exit of the tube being at atmospheric pressure. The mass flow rates of the air and the water are 7.9×10^{-3} kgs^{-1} and 0.35 kgs^{-1}, respectively. Determine the frictional pressure drop along the tube for the two-phase flow. The density of the gas is 1.18 kgm^{-3}, and the viscosity is 1.85×10^{-5} Nsm^{-2}, while the density of the liquid is 997 kgm^{-3} and the viscosity is 8.9×10^{-4} Nsm^{-2}.

Solution

The overall pressure drop or pressure gradient along an inclined pipe is the sum of the frictional, gravitational, and acceleration pressure gradients:

$$\frac{dp}{dz} = \left(\frac{dp}{dz}\right)_f + \left(\frac{dp}{dz}\right)_{grav} + \left(\frac{dp}{dz}\right)_a \tag{6.60}$$

A momentum balance is therefore

$$\frac{dp}{dz} = \frac{P\tau_w}{A} + \rho g \sin\theta + \rho U \frac{dU}{dz} \tag{6.61}$$

The most widely used frictional pressure drop calculations for two-phase flow are historically based on calculations that use pressure drop multipliers. The most widely used correlation that uses multipliers is that of Lockhart and Martinelli (1949), where

$$\left(\frac{dp}{dz}\right)_f = \varphi_g^2 \left(\frac{dp_L}{dz}\right)_g = \varphi_L^2 \left(\frac{dp_L}{dz}\right)_L \tag{6.62}$$

The multipliers φ_g and φ_L are a function of the parameter X defined as

$$X^2 = \frac{\left(\dfrac{dp_L}{dz}\right)_L}{\left(\dfrac{dp_L}{dz}\right)_g} = \frac{\varphi_g^2}{\varphi_L^2} \tag{6.63}$$

The multipliers are dependent on the type of flow as either laminar (referred to as viscous) or turbulent. The liquid pressure drop multiplier is alternatively given as

$$\varphi_L^2 = 1 + \frac{C}{X} + \frac{1}{X^2} \tag{6.64}$$

where C is a dimensionless parameter and depends on the nature of the flow of both the gas and liquid given in Table 6.1.

For the gas flow only, the mass flow per unit flow area, or mass loading, is

$$G_g = \frac{\dot{m}_g}{A} = \frac{4\dot{m}_g}{\pi d^2} = \frac{4 \times 7.9 \times 10^{-3}}{\pi \times 0.03^2} = 11.18 \text{ kgm}^{-2}\text{s}^{-1} \tag{6.65}$$

For which the Reynolds number in the tube is

$$Re_g = \frac{G_g d}{\mu_g} = \frac{11.18 \times 0.03}{1.85 \times 10^{-5}} = 18,125 \tag{6.66}$$

TABLE 6.1

Dimensionless Parameter
for Pressure Drop Multiplier

Liquid (L)	Gas (G)	C
Turbulent	Turbulent	20
Laminar	Turbulent	12
Turbulent	Laminar	10
Laminar	Laminar	5

This corresponds to the turbulent flow for which a suitable friction factor correlation is

$$\frac{1}{\sqrt{f_g}} = -3.6 \times \log_{10}\left(\frac{6.9}{Re_g}\right) = -3.6 \times \log_{10}\left(\frac{6.9}{18,125}\right) \tag{6.67}$$

This gives a friction factor of 0.0066. The pressure drop of the air per unit metre length along the tube is therefore

$$\left(\frac{dp}{dz}\right)_g = \frac{2f_g \rho_g U_g^2}{d} = \frac{2f_g G_g^2}{\rho_g d} = \frac{2 \times 0.0066 \times 11.18^2}{1.18 \times 0.03} = 46.6 \ \text{Nm}^{-2}\text{m}^{-1} \tag{6.68}$$

For the water, the mass loading (mass flow per unit area) is

$$G_L = \frac{\dot{m}_L}{A} = \frac{4\dot{m}_L}{\pi d^2} = \frac{4 \times 0.35}{\pi \times 0.03^2} = 495.1 \ \text{kgm}^{-2}\text{s}^{-1} \tag{6.69}$$

The corresponding Reynolds number in the tube is

$$Re_L = \frac{G_L d}{\mu_L} = \frac{495.1 \times 0.03}{8.9 \times 10^{-4}} = 16,689 \tag{6.70}$$

This also corresponds to turbulent flow for which the friction factor is

$$\frac{1}{\sqrt{f_L}} = -3.6 \times \log_{10}\left(\frac{6.9}{Re_L}\right) = -3.6 \times \log_{10}\left(\frac{6.9}{16,689}\right) \tag{6.71}$$

This gives a friction factor of 0.0067. The liquid pressure drop per metre length is

$$\left(\frac{dp}{dz}\right)_L = \frac{2f_L \rho_L U_L^2}{d} = \frac{2f_L G_L^2}{\rho_L d} = \frac{2 \times 0.0067 \times 495.1^2}{997 \times 0.03} = 110 \ \text{Nm}^{-2}\text{m}^{-1} \tag{6.72}$$

The Lockhart-Martinelli parameter, X, is given by

$$X = \sqrt{\frac{\left(\dfrac{dp}{dz}\right)_L}{\left(\dfrac{dp}{dz}\right)_g}} = \sqrt{\frac{110}{46.6}} = 1.54 \qquad (6.73)$$

Since both phases are in the turbulent phase, the two-phase multiplier is found from Equation 6.64 and Table 6.1 to be

$$\varphi_L^2 = 1 + \frac{20}{1.54} + \frac{1}{1.54^2} = 14.17 \qquad (6.74)$$

The combined gas and liquid pressure drop per metre length is therefore

$$\frac{dp}{dz} = \varphi_L^2 \left(\frac{dp}{dz}\right)_L = 14.17 \times 110 = 1558 \ \mathrm{Nm^{-2}m^{-1}} \qquad (6.75)$$

The pressure drop for the two-phase mixture for the total length of pipe of 20 m is therefore 31.2 kNm⁻².

What Should We Look Out For?

In spite of its continued widespread use, the Lockhart-Martinelli correlation does not predict the effect of mass flux as adequately well as some other parameters. A number of more sophisticated correlations have subsequently been produced as replacements and may also be worth considering.

What Else Is Interesting?

A good correlation to determine the gas void fraction is

$$\alpha = 1 - \frac{1}{\sqrt{1 + \dfrac{20}{X} + \dfrac{1}{X^2}}} \qquad (6.76)$$

The liquid hold-up is the fraction of pipe volume occupied by the liquid and can be found using the Martinelli parameter (Figure 6.4):

$$\lambda_L = 1 - \left(1 + X^{0.8}\right)^{-0.378} \qquad (6.77)$$

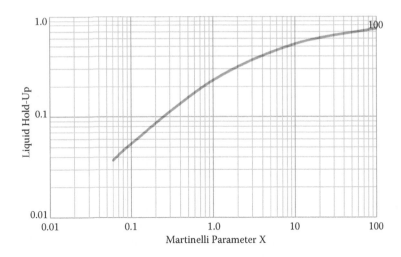

FIGURE 6.4
Liquid Hold-Up

In experimental testing where practical, the liquid hold-up can be measured by trapping flow segments between quick-closing valves and measuring the entrapped liquid.

Problem 6.11: Flow in Horizontal Pipes

An experimental test apparatus is used to examine two-phase flow regimes in horizontal pipelines. A particular experiment involves the flow of air and water which flow through a horizontal glass tube with a length of 40 m and an internal diameter of 25.4 mm. The water flows at a controlled rate of 0.026 kgs^{-1} at one end with air at a controlled rate of 5 × 10^{-4} kgs^{-1} injected in the same direction. If the density of water is 1000 kgm^{-3}, and the density of air is 1.2 kgm^{-3} at the test temperature, determine the total mass flux, the mean density of the mixture, and the flow regime.

Solution

The air and water superficial velocities can be determined from the mass flow rates:

$$j_g = \frac{Q_g}{A} = \frac{4\dot{m}_g}{\rho \pi d^2} = \frac{4 \times 0.0005}{1.2 \times \pi \times 0.0254^2} = 0.822 \text{ ms}^{-1}$$

$$j_L = \frac{Q_L}{A} = \frac{4\dot{m}_L}{\rho \pi d^2} = \frac{4 \times 0.026}{1000 \times \pi \times 0.0254^2} = 0.0513 \text{ ms}^{-1}$$

(6.78)

The total mass flux through the glass tube is therefore

$$G = \rho_g j_g + \rho_L j_L = 1.2 \times 0.822 + 1000 \times 0.0513 = 52.3 \text{ kgm}^{-2}\text{s}^{-1} \quad (6.79)$$

The mass flux can alternatively be found from the total mass flow rate, which is the combined flow rate of both the air and water (0.0265 kgs⁻¹) and the flow area:

$$G = \frac{\dot{m}}{A} = \frac{4 \times 0.0265}{\pi \times 0.0254^2} = 52.3 \text{ kgm}^{-2}\text{s}^{-1} \quad (6.80)$$

Using the modified Baker plot, the two-phase flow regime for the flow mixture is found using the dimensionless correction factors. These were originally developed to account for variations in the density, viscosity, and surface tension of the flowing media and are functions of the fluid properties and normalised with respect to the properties of water and air at standard conditions. At standard conditions, both factors are equal to 1:

$$\lambda = \left(\frac{\rho_g \rho_L}{\rho_a \rho_w} \right)^{\frac{1}{2}} \quad (6.81)$$

and

$$\Psi = \frac{\sigma_w}{\sigma_L} \left(\frac{\mu_L}{\mu_w} \left(\frac{\rho_w}{\rho_L} \right)^2 \right)^{\frac{1}{3}} \quad (6.82)$$

The viscosity of water is 0.001 Nsm⁻², and the surface tension of air and water is 0.073 Nm⁻¹. Therefore,

$$\lambda = 1.0$$

$$\Psi = 1.0$$

$$G_g = \rho_g j_g = 1.2 \times 0.822 \approx 1.0 \quad (6.83)$$

$$G_L = \rho_L j_L = 1000 \times 0.0513 = 51.3$$

This corresponds to stratified flow (Figure 6.5). It is worth noting that by increasing the water mass flow rate, slug/plug flow can be achieved. In this case,

$$G_L \Psi = 100 = \rho_L j_L \Psi = 1000 \times j_L \times 1 \quad (6.84)$$

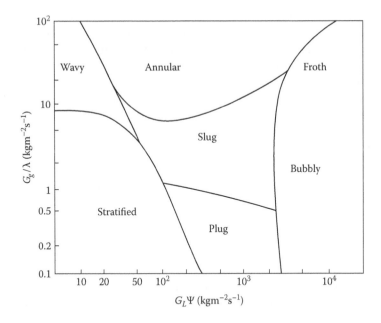

FIGURE 6.5
Modified Baker Plot for Horizontal Flow

The superficial velocity of the water is therefore

$$j_L = 0.1 = \frac{4\dot{m}_L}{\rho_L \pi d^2} = \frac{4 \times \dot{m}_L}{1000 \times \pi \times 0.0254^2} \tag{6.85}$$

Solving gives a mass flow rate of liquid as 0.05 kgs⁻¹.

What Should We Look Out For?

The Baker plot, which is also known as the Baker map, is reasonable for water and air mixtures as well as oil and gas mixtures in pipes with diameters less than 5 cm. A degree of caution should, however, be used when using multi-phase mixtures which deviate radically from water and air and for larger-diameter pipes. There are a number of other two-phase flow regime maps that have been developed for horizontal and vertical flows. The Taitel and Dukler map produced in 1976 is another of the most widely used flow maps for two-phase flow and is based on a semi-theoretical method. It also uses the Lockhart-Martinelli parameter but is generally regarded as being more difficult to use.

What Else Is Interesting?

The flow of two-phase mixtures under certain conditions is known to be responsible for causing erosion of pipework and in particular sub-sea pipelines carrying crude oil. Pipelines carrying heavy wax crudes can result in wax deposition, which can be controlled by maintaining a high temperature or by using pigging or inhibitors. Pigging involves using a device to clean or clear away the inside of a pipeline carrying natural gas and crude oil, which are prone to the build-up of deposits such as waxes and hydrates. They are made of rubber or polyurethane, and the basic design consists of two plates held apart by a short rod. They can also incorporate various sensing and recording equipment. They are launched into the pipeline and move under the effect of an applied pressure. They emit a squealing sound as their blades scrape along the pipeline and are recovered from a receiving trap, which are loops in the pipeline that can be isolated by shut-off valves.

For fluids with high velocities, erosion of pipework may occur according to the semi-empirical equation:

$$\rho U^2 = 15,000 \tag{6.86}$$

and is used to determine the maximum allowable velocity to avoid the damaging effects. For a two-phase mixture, the average density of the two-phase mixture and superficial velocity is

$$\rho j^2 = \left(\rho_g \alpha + \rho_L (1 - \alpha) \right) \left(j_g + j_L \right)^2 \tag{6.87}$$

For the erosion criteria, this is therefore

$$\left(j_g^2 + 2 j_g j_L + j_L^2 \right) = \frac{15,000}{\left(\rho_g \alpha + \rho_L (1 - \alpha) \right)} \tag{6.88}$$

The quadratic can be solved and expressed in terms of either superficial velocity.

Unlike erosion, corrosion of pipelines is the unwanted wastage of metallic materials due to reaction with the environment. The effect includes the loss of strength of material, a change in appearance, change in surface heat transfer and fluid flow properties, contamination, seizure, electrical contact failure, leakage, and general surface damage. Corrosion rates are determined to a large extent by the chemical nature of the process stream and its pressure and temperature; due account must be taken of the flow conditions and how they interact with the ongoing chemical processes. In pipelines carrying oil with water at low velocities, the water can separate to form a layer below the oil, and depending on the oil may contain carbon dioxide and hydrogen sulphide, resulting in corrosion to the pipe.

Problem 6.12: Bubbly Flow

Air is injected through a series of holes on the underside of a large cylindrical tank containing water. The holes have a diameter of 0.5 mm, for which the equivalent diameter of the bubbles which form at the holes is 1 mm. Determine the rise velocity of the air bubble in the tank using the Reynolds-Eötvös-Morton diagram (Figure 6.6). The density of the water is 1000 kgm⁻³, and viscosity is 0.001 Nsm⁻². The surface tension is 0.07 Nm⁻¹. The density of gas is 1.1 kgm⁻³.

Solution

This problem may be simplified by assuming that the forming bubbles do not coalesce or break up into smaller bubbles. The bubble shape can be found from the graph using the dimensionless Morton and Eötvös numbers. The Morton number is

$$Mo = \frac{(\rho_L - \rho_g)\mu_L g}{\rho_L^2 \sigma^2} = \frac{(1000 - 1.1) \times 0.001}{1000^2 \times 0.07^2} = 2.86 \times 10^{11} \qquad (6.89)$$

The Eötvös number is

$$Eo = \frac{g(\rho_L - \rho_g)d_b^2}{\sigma} = \frac{9.81 \times (1000 - 1.1) \times 0.001^2}{0.07} = 0.14 \qquad (6.90)$$

From Figure 6.6, $\log_{10}(Mo)$ is –10.54; the Reynolds number is found to be 130. The velocity of the bubbles is therefore

$$U_b = \frac{Re\mu_L}{\rho_L d_b} = \frac{130 \times 0.001}{1000 \times 0.001} = 0.13 \text{ ms}^{-1} \qquad (6.91)$$

What Should We Look Out For?

The use of small holes ensures that there is a high surface area which is necessary where sparging is used to transfer gas into a liquid. Larger holes may lead to coalescence and rapid velocity reducing the effectiveness of mass transfer.

What Else Is Interesting?

The Eötvös number may be seen as being proportional to buoyancy force divided by surface tension force. It is named after Hungarian physicist Loránd Eötvös (1848–1919).

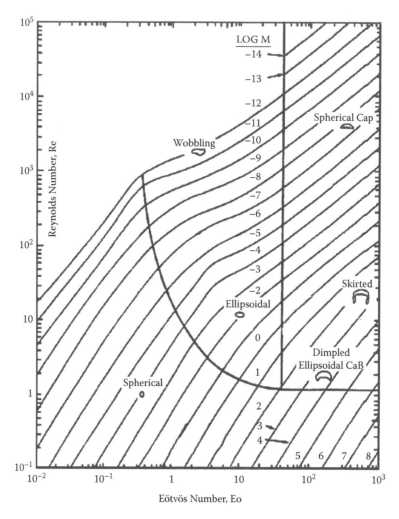

FIGURE 6.6
Reynolds-Eötvös-Morton Diagram (Reprinted from Ansari, M.R. and Nimvari, M.E., 2011, Bubble Viscosity Effect on Internal Circulation within the Bubble Rising Due to Buoyancy Using the Level Set Method, *Annals of Nuclear Energy* 38: 2770–2778, with permission from Elsevier.)

Further Problems

1. Determine the range of mean density of a mixture of air in a 50:50 oil–water liquid phase across a range of gas void fractions. The density of oil is 900 kgm⁻³, water is 1000 kgm⁻³, and gas is 10 kgm⁻³.

2. Describe, with the use of sketches, the various flow regimes that can exist in a vertical pipe carrying two-phase flow (liquid and gas).

3. A mixture of air and water at a temperature of 25°C flows up through a vertical tube with a length of 4 m and an internal diameter of 25.4 mm with the exit of the tube being at atmospheric pressure. The mass flows of the air and the water are 0.007 kgs^{-1} and 0.3 kgs^{-1}, respectively. For air, the density is 1.2 kgm^{-3} and viscosity is 1.85 × 10^{-5} Nsm^{-2}, and for water, the density is 1000 kgm^{-3} and viscosity is 8.9 × 10^{-4} Nsm^{-2}. *Answer:* 2.7 kNm^{-2}m^{-1}

4. An experimental test rig is used to examine two-phase flow regimes in horizontal pipelines. A particular experiment involved uses air and water at a temperature of 25°C, which flow through a horizontal glass tube with an internal diameter of 25.4 mm and a length of 40 m. Water is admitted at a controlled rate of 0.026 kgs^{-1} at one end and air at a rate of 5 × 10^{-4} kgs^{-1} in the same direction. The density of water is 1000 kgm^{-3}, and the density of air is 1.2 kgm^{-3}. Determine the mass flow rate, the mean density, gas void fraction, and the superficial velocities of the air and water. *Answer:* 0.02605 kgs^{-1}, 61.1 kgm^{-3}, 0.94, 0.822 ms^{-1}, 0.051 ms^{-1}

5. Describe, with the use of sketches, the various two-phase flow regimes that can exist in a horizontal pipe carrying a liquid and a gas.

6. Explain what is meant by gas hold-up and describe ways in which it can be measured.

7. It is required to transport a hydrocarbon as a two-phase mixture of liquid and vapour along a smooth-walled pipe with an inside diameter of 100 mm. The total hydrocarbon flow rate is 2.4 kgs^{-1} with a vapour mass fraction of 0.085. The pipe is to operate at an absolute pressure of 2.2 bar. The liquid density is 720 kgm^{-3}, and viscosity is 4.8 × 10^{-4} Nsm^{-2}, while for the vapour, the density is 1.63 kgm^{-3}, and the viscosity is 2.7 × 10^{-5} Nsm^{-2}. Determine the maximum permissible length of pipe if the pressure drop along the pipe is not to exceed 20 kNm^{-2}. *Answer:* 44 m

8. A gas is admitted at a rate of 0.015 m^{3}s^{-1} to a vertical glass pipe with an inside diameter of 50 mm. The gas bubbles that form travel with a velocity of 32 ms^{-1}. Determine the gas void fraction and the velocity of the liquid if the volumetric flow is 2.5 × 10^{-5} m^{3}s^{-1}. *Answer:* 0.24, 1.7 ms^{-1}

9. Characterise the main concepts of a homogeneous flow model, separated flow models, and specific flow pattern models.

10. A mixture of high pressure water and steam at a rate of 0.5 kgs^{-1} flows up a vertical tube with an inside diameter of 25.4 mm at a pressure 22 bar. Determine the type of flow if the mass quality is 1%. The density of the water is 845 kgm^{-3}, the density of steam is 10.8 kgm^{-3}, and the viscosity of the water is 1.24 × 10^{-4} Nsm^{-2}. *Answer:* Slug flow

11. Define the terms *slip velocity* and *liquid hold-up* in two-phase pipe flow of a gas–vapour mixture.

12. A mixture of oil and gas flows through a horizontal pipe with an inside diameter of 150 mm. The respective volumetric flow rates for the oil and gas are 0.015 and 0.29 m^3s^{-1}. Determine the gas void fraction and the average velocities of the oil and gas. The friction factor may be assumed to be 0.0045. The gas has a density of 2.4 kgm^{-3} and viscosity of 1×10^{-5} Nsm^{-2}. The oil has a density of 810 kgm^{-3} and density of 0.82 Nsm^{-2}. *Answer*: 0.79, 20.8 ms^{-1}, 4 ms^{-1}

13. Show that the gas void fraction for a flowing gas–liquid mixture can be expressed in terms of the phase velocity, quality, and densities of the mixture as

$$\alpha = \frac{1}{1 + \dfrac{\rho_g U_g (1-x)}{\rho_L U_L x}}$$

14. A bubbly mixture of gas and liquid flows up a vertical glass tube with an internal diameter of 25 mm. The liquid flow is controlled to be 0.02 litres per second, and the gas flow is 10 litres per second. The bubble velocity is determined photographically to have a velocity of 30 ms^{-1}. Determine the gas void fraction for the two-phase mixture and the liquid velocity. *Answer*: 0.68, 0.13 ms^{-1}

15. Show that for a one-dimensional annular flow in a horizontal pipe with no acceleration, the pressure gradient on the gas core is

$$\frac{dp}{dz} = \frac{4\tau_i}{d\sqrt{\alpha}}$$

where τ_i is the interfacial shear stress and α is the gas void fraction.

7

Fluid Mixing

Introduction

Mixing is the process of bringing two or more components and/or phases together into intimate contact in order to achieve a desired outcome. Typically used to promote homogeneity, promote the dissolution of a solute, or enhance the rate of mass transfer or the rate of a chemical reaction, mixing may involve dispersing one phase through. Dispersing a gas in the form of bubbles through liquids such as through the action of sparging or through fluidized beds of solid particulates may enhance both heat and mass transfer. Alternatively, solid particles can be dispersed through a liquid for the same purpose.

In its simplest form, the process of mixing consists of a vessel within which the various materials to be dispersed are contained (Figure 7.1). The materials may range from dry, free-flowing solids through to thick viscous slurries, pastes, and doughs. The movement and flow of the materials to cause dispersion involves many complex interactions and may be time dependent and further complicated by physical, chemical, or biochemical reactions that may also simultaneously take place. The effectiveness of mixing may be further enhanced by the use of mechanical devices such as rotating agitators, impellers, and propellers. These consist of a rotating shaft upon which blades are attached to cause either radial or axial flow. Radial-flow impellers discharge the fluid in a horizontal or radial direction from the rotating shaft, whereas propellers provide axial flow. Downward axial flow produces strong top-to-bottom currents. The presence of baffles attached to the wall of the vessel can prevent the undesirable effect of swirling or vortexing by interrupting the flow patterns.

Mixing is a widely practiced operation in the food industry and is used for blending such as for teas and coffees, in the production of dried soups, cake mixes, and custard powders, and for producing emulsions such as ice cream mixes. The mixing of solids and liquids is used for canned foods, dairy products, and chocolates and confectionary in which ingredients are mixed in a more or less liquid state and solidify upon cooling. Blending is a form of mixing that involves particulates or powders and is the combination

FIGURE 7.1
Mixing Vessel with a Control Panel (Courtesy of David Thomson.)

of different materials and their spatial distribution and is carried out until a required degree of homogeneity has been achieved.

In a mixing process that requires the dispersion of a solid in a liquid in which a chemical reaction also takes place between the two phases, it may be the case that the reaction may occur only when a suitable degree of dispersion has been achieved. Likewise, for a mixing process that involves the dispersion of a gas through a liquid, a chemical reaction may take place on account of the degree of dispersion. It does not necessarily follow, however, that the dispersion of one phase in another would produce a reliable design. This is because there are many definitions of the term *dispersion,* and there are mixing variables that provide an effective dispersion of the phases that may correlate oppositely with reference to mass transfer effects.

A mixing process that involves complex flow patterns with a chemical reaction would normally be expressed in terms of the required mass transfer rate. Invariably, mass transfer is not directly related to dispersion. Determining the correct reaction conditions requires additional information concerning the concentration of the solid and liquid phases, along with other aspects pertinent to the quantitative performance of the mixing process.

Problem 7.1: Liquid–Solid Mixing

A vessel contains water in which insoluble solid particles are required to be kept in suspension with the use of a stirred agitator. Identify the difficulties of operation to ensure that the particles remain in suspension.

Solution

The suspension of solids is one of the most common applications used in mixing. The suspension of solids can be classified as having either on-bottom motion, which refers to particles moving with some velocity on the bottom of the tank, or off-bottom motion in which particles move off the bottom with some vertical velocity. Typically, finer particles may be almost uniformly suspended throughout the tank while coarser particles barely move off the bottom of the tank.

In general, solid particles with low settling velocities of up to 150 mm per minute are suspended uniformly and without difficulty. The free-settling of solids having rates of settling velocities of between 8 and 100 mm per second require suspension, whereas hindered-settling particles settle so slowly that they are treated not as a solids suspension but as blending or fluid motion.

The distribution of suspended solid particles during mixing is dependent on many factors. Complete uniformity by agitation can be achieved for a given practical power input. Vertical-settling particles are suspended to the same degree at the top of the tank as they are throughout the rest.

What Should We Look Out For?

Power failure can cause major mechanical problems in liquid–solids mixing systems. Depending upon the depth of the solids and the degree of packing of the solid particles, the blades of the mixer may become completely buried. Start-up may then be impossible, given normal motor and gear reducer ratings. Considerable damage can therefore be caused to the mixer if the motors are repeatedly jigged and the mixer fails to rotate. The resulting inertial load can therefore cause excessive instantaneous loads on the gear mechanism. The start-up of equipment with settled solids therefore requires careful studies to be completed to determine the required torque for rotation. The gearbox and shaft design must then be altered accordingly. Alternatively, a liquid or gas lance can be used to suspend the particles around the mixer blades. This device is inserted into the solid mass, and a gas or liquid under pressure then loosens the packed solids in the mixing zone, thereby freeing the blades of the mixer.

What Else Is Interesting?

There is a maximum solids concentration for a given mixing system, beyond which the fluid motion stops and any reasonable increase in applied power does not restore motion. This maximum concentration is known as the ultimate weight percent settled solids and is the percentage by weight of solids that settle until no further supernatant liquid collects above them. In other words, it is the highest solids concentration that still allows liquid to surround the particles, which may typically be in the range of 70% to 90%.

While coarse material will not settle out, the disadvantage is that the mixer may be required to move a highly viscous non-Newtonian form of slurry. The viscosity of the slurry will largely depend upon the particle size such that a slurry of fine particles will require less power at lower percent solids and require higher power at higher percent solids than coarse particles.

The actual power drawn by an impeller for rotation depends upon the uniformity of the slurry. If the solids have settled below the impeller, then once it is turned on it is influenced only on the density of the liquid. If the solids are tightly packed, a new tank bottom is, in effect, created with the new bottom effectively being closer to the impeller, which changes the power requirement. As solids are picked up off the bottom, the density of the slurry then influences the operation of the impeller. During off-bottom suspension conditions, the concentration of the slurry around the impeller is higher than the average concentration when the solids are fully suspended. This draws more power than at suspension equilibrium.

Not all mixing operations involve stirrers and agitators. In-line mixers are static devices with no moving parts and use pipelines that carry fluids for mixing, in which fluid streamlines are caused to cross one another, thereby intimately mixing the fluid. They are used to mix fluids such as dispersing gases in liquids and solids in liquid, and assist with dissolving solids in liquids and mixing immiscible liquids. They consist of a twisted ribbon of chemically inert and unreactive material such as metal or plastic of sufficient length and are firmly held in place to provide good mixing characteristics. They are also known as static mixers.

The effectiveness of mixing can be predicted using computational fluid dynamics (CFDs). For example, in the case of shell and tube heat exchangers, baffles are often used to promote effective heat exchange. A CFD simulation can be used to examine the performance of flow and heat exchange as shown in Figure 7.2. The shaded areas illustrate the high and low velocities as the fluid moves from compartment to compartment formed by baffles.

Where neither real-time testing nor CFDs are able to be used due to complexities of the fluid or the process, the performance of mixing can, when possible, be determined experimentally. The use of tracer materials, for example, involves tracking the dispersion of extraneous materials by various

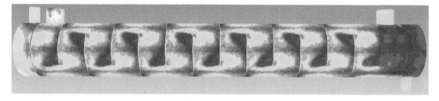

FIGURE 7.2
CFD Simulation of a Shell-and-Tube Heat Exchanger

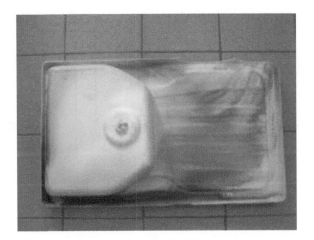

FIGURE 7.3
Flow and Mixing Profiles of Materials in Kitchen Sink Manufacturing (Photo from C.J. Schaschke.)

detection or analytical techniques. An example used to determine the dispersion and mixing of polymeric materials is illustrated in Figure 7.3 as used in the production process of synthetic polymer-based kitchen sinks. Materials are prepared by mixing and injecting into moulds for curing or hardening of the polymer. To ensure the necessary dispersion of materials, tests are carried out involving dye patches on the surface. These capture the flow direction and the effective mixing of the materials.

Problem 7.2: Connected Mixing Tanks

A process involves preparing a salt solution using three connected continuously stirred tanks. Each tank contains 1000 litres of water in which tank A flows into B, which flows into C, which in turn flows back into A. The flow between tanks is maintained at 50 litres per minute. If tanks A and C initially contain a solution of salt with concentrations of 3 and 6 grams per litre, respectively, while tank B vessel contains only water, determine the time for tank B to reach a final concentration of 2.9 g L^{-1}.

Solution

The three tanks are connected as shown in Figure 7.4. Assuming that the tanks are perfectly mixed and the concentration is uniform throughout each tank, the flow to and from each tank is the same. The change in concentration in each tank can be determined from a mass balance for each tank as

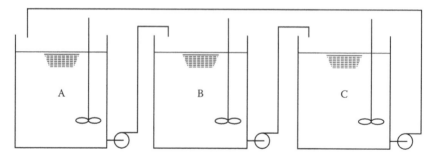

FIGURE 7.4
Connected Tanks

$$\dot{m}C_{in} = V\frac{dC}{dt} + \dot{m}C_{out} \tag{7.1}$$

For each tank, the rate of change of concentration is therefore

$$\frac{dC_A}{dt} = \frac{\dot{m}}{V}(C_C - C_A)$$

$$\frac{dC_B}{dt} = \frac{\dot{m}}{V}(C_A - C_B) \tag{7.2}$$

$$\frac{dC_C}{dt} = \frac{\dot{m}}{V}(C_B - C_C)$$

The three equations can be solved simultaneously using a simple algorithm using a small increment of time, say 0.01 minutes, starting with the initial concentrations of

$$C_{Ao} = 3.0gL^{-1}$$

$$C_{Bo} = 0.0gL^{-1} \tag{7.3}$$

$$C_{Co} = 6.0gL^{-1}$$

After each incremental period of time, the new concentration is calculated. Concentration profiles of the three tanks are shown in Figure 7.5.

The eventual final concentration is 3 gL^{-1} for each of the three tanks. The concentration in tank C falls from 6 gL^{-1} to its final concentration and feeds into tank A, which as a consequence, gains in concentration but then falls as it is continuously washed out into B. The concentration of tank B rises until the time reach at 2.9 gL^{-1} corresponds to 0.91 minutes (55 seconds).

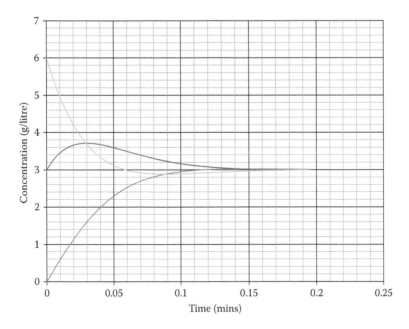

FIGURE 7.5
Concentration Profiles in Connecting Tanks

What Should We Look Out For?

This problem is solved using a material balance for each tank. A material balance is the exact accounting of all material that enters, leaves, accumulates, or is depleted within a given period of operation across an imaginary boundary of a process or part of a process. It is, in effect, an expression of the law of conservation of mass in accounting terms. Where the conditions of a process are steady and unvarying with time, the material input flow across the boundary equals the output flow. Depending on whether a chemical or physical change occurs, material balances may be conveniently calculated and expressed in either mass or molar terms.

What Else Is Interesting?

A continuous stirred-tank reactor (CSTR) is a common type of reactor vessel which is stirred in such a way that it may be assumed that the contents are perfectly mixed. The output composition of the material is therefore also assumed to be the same composition at all points within the reactor. This assumption is convenient for simple cases, but this is often not the case in practice. Therefore, tests are required to be carried out to determine the

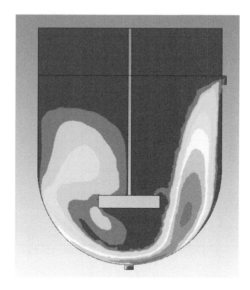

FIGURE 7.6
CFD Simulation of a CSTR

ability of a mixer to reach homogeneity, which may depend on the type of mixer, its location in the tank, its rotational speed, and the properties of the fluid being mixed. Computation fluid dynamics can be used to simulate the effectiveness of mixing, for which a typical example is shown in Figure 7.6.

Problem 7.3: Continuously Stirred Tanks in Series

A process consists of a series of five continuously stirred tanks (CSTs) connected in series. Each tank has a capacity of 50 litres and contains water. If the first tank contains a salt solution initially at a concentration of 20 gL^{-1}, determine the concentration in each of the five tanks over time.

Solution

The five CSTs are connected as shown in Figure 7.7. The solutions in each of the tanks are assumed to be perfectly mixed, and the concentration is uniform throughout each tank. As the salt is washed from the first tank, the salt concentration will rise in the second until it, too, is washed out, while at the same time, some salt is washed to the third tank, and so on. This can be described using a mass balance for each tank (Equation 7.1). For each tank, the rate of change of concentration is therefore

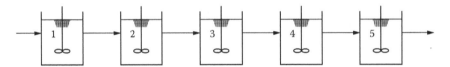

FIGURE 7.7
CSTs in Series

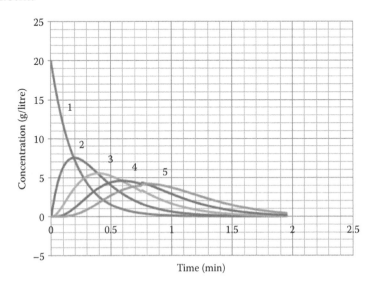

FIGURE 7.8
Concentration Profiles in Connecting Tanks

$$\frac{dC_n}{dt} = \frac{\dot{m}}{V}(C_{n-1} - C_n) \tag{7.4}$$

The concentration of salt in each tank can be determined simultaneously with a simple algorithm using a small increment of time, say 0.01 minute, starting with the initial concentrations of 20 gL^{-1} of salt in the first tank. Concentration profiles of the five tanks are shown in Figure 7.8.

What Should We Look Out For?

While the concentration of the salt in the first tank rapidly decays, the appearance of salt in the subsequent tanks rises and falls as it, too, is washed out. In this problem, the tanks are used for dilution. Such tanks are often used as chemical reactors in which controlled chemical reactions take place. Where many such reaction tanks are connected in series, the concentration profile approximates that of a single plug flow reactor (PFR). This is a type of idealized tubular reactor that features no radial or axial mixing. All of the components that flow through the reactor therefore possess the same residence time. A PFR may consist of either one long reactor or many short reactors as a tube bundle. The reactants flow through the length of the reactor and during this process the chemical reaction rate changes. They are usually used for gas-phase reactions that require high temperatures. Used for both exothermic and endothermic reactions, heat transfer is effective through the tube walls.

What Else Is Interesting?

Many mixing operations involve a batch vessel in which the vessel is charged and mixed by way of a mechanical agitator until a desired outcome has been achieved. In the case of mixing paints, for example, the objective may be to achieve a uniform colour, or with foods it may be to achieve a certain consistency. Batch mixing generally requires around 10% to 25% more power for stable conditions, compared with continuous mixing operations. When designing a continuous system, provision must be made in case there is a power failure that would stop the mixer or interrupt the feed or discharge streams. An auxiliary recirculation pump may be used to prevent solids from settling. In a continuous flow, the entire tank need not have the same composition as the feed and exit streams. The only process requirement is for the composition at the draw-off point to be constant. The mixer should therefore be large enough to maintain the composition at this point, and the rest of the tank can have different concentrations than those at the draw-off point. Applying successively larger amounts of mixer power will make the tank contents uniform.

Problem 7.4: Dimensional Analysis

Derive the relationship between the power required for mixing by way of an impeller, the physical properties of the liquid, and the physical dimensions of the vessel within which it is contained.

Solution

The relationship between the power required for mixing and the physical nature of the liquid and vessel can be derived using dimensional analysis. Dimensional analysis is a useful tool for checking an expression or a solution to a problem that is used to describe an observable phenomenon. It requires dimensional consistency between the variables and the phenomenon. The relationships of the variables are considered in terms of their fundamental dimensions and are regrouped into dimensionless groups. These groups, by virtue that they have no scale of size, mass, or time, can then be used to study the effects of any terms in any situation independent of its application. Dimensionless groups are therefore widely used in engineering and are particularly important in both scale-up and scale-down processes.

In this case, for liquids that show Newtonian tendencies, the power for mixing can be expressed in terms of the density and viscosity of the liquid being mixed, as well as the diameter of the vessel within which mixing takes place, and the rotational speed of the mixer. The gravitational acceleration is also important here. The power can therefore be presented as a function of these variables in a power relationship as

$$P_o = kD^a \mu^b g^c \rho^d N^e \tag{7.5}$$

Expressed in terms of fundamental dimensions of mass, length, and time, this is

$$ML^2T^{-3} = (L)^a \left(ML^{-1}T^{-1}\right)^b \left(LT^{-2}\right)^c \left(ML^{-3}\right)^d \left(T^{-1}\right)^e \tag{7.6}$$

The exponents a to e can be expressed in terms of b and c as

$$a = 5 - 2b - c$$

$$d = 1 - b \tag{7.7}$$

$$e = 3 - b - 2c$$

For which the following three dimensionless groupings formed are:

$$\frac{P_o}{\rho N^3 D^5} = k\left(\frac{D^2 N \rho}{\mu}\right)^{-b}\left(\frac{DN^2}{g}\right)^{-c} \tag{7.8}$$

These dimensionless groups are presented as

$$N_P = k\,\mathrm{Re}^{-b}\,Fr^{-c} \tag{7.9}$$

where N_P is the power number, Re is a form of a Reynolds number, and *Fr* is known as the Froude number, which represents the influence of gravitation. There are a number of published correlations between the power and Froude number for baffled and unbaffled vessels as well as different types of stirrers.

What Should We Look Out For?

In dimensional analysis, it is essential that only the relevant independent variables be correctly identified. The inclusion of irrelevant variables will not result in a meaningful relationship. The number of dimensional groups that will be formed is equal to the difference between the number of variables including the dependent variable and the number of dimensions.

What Else Is Interesting?

The Reynolds number is named after the British engineer Osborne Reynolds (1842–1912). It expresses the ratio of inertial to viscous forces in a flowing fluid. More commonly used to determine the flow regime in a pipe of circular cross section, there are many other forms of the Reynolds number. The Froude number named after the British engineer William Froude (1810–1879), who is noted for his work on the laws for the resistance of the hulls of ships in water and in predicting their stability. The Froude number is a widely used dimensionless group that represents the influence of gravity in the power relationship for fluid systems such as pumping, mixing in unbaffled tanks and reactors, and determining the extent of fluidization of particles in a fluidized bed.

Problem 7.5: Impeller Power Requirement for Mixing

A bioreactor with an internal diameter of 2 m is equipped with baffles and two sets of flat blade impellers and four baffles. The impeller has a diameter of 1.25 m, and the baffle width is 20 cm. The viscosity of the liquid broth is 0.025 Nsm^{-2} and the density is 1220 kgm^{-3}. Determine the power number and the number of sets of impellers required if the rotational speed of the impeller is 60 rpm.

Solution

A bioreactor is a vessel used for biological processing containing the growth of living cells or tissues either as the product themselves or as biocatalysts in the production of other products. There are many designs of bioreactors. The most common are cylindrical and mounted on their axis and range in capacity from a few litres to many cubic metres. Small bioreactors are made of glass, while large bioreactors are fabricated from stainless steel. The mode of operation is batch, fed batch, or continuous. Constant agitation is usually maintained with an appropriate stirrer that also aids in oxygen transfer in aerobic processes.

For the bioreactor, the modified Reynolds number is

$$Re = \frac{\rho N^2 D}{\mu} = \frac{1220 \times (60/60)^2 \times 1.2}{0.025} = 58,560 \qquad (7.10)$$

For a vessel in which the baffle corresponds to 10% of the diameter, the power number is 6. The power required for mixing is therefore

$$P_o = \rho N^3 D^5 N_P = 1220 \times (60/60)^3 \times 1.2^5 \times 6 = 18,214 \text{ W} \qquad (7.11)$$

The power required using two sets of impellers is 36.4 kW.

What Should We Look Out For?

Bioreactors are used to contain biological processes that may be in the form of a suspension of cells, immobilized cells, or enzymes, and depending on the living organism, operated aerobically or anaerobically. The process materials are often fragile in nature and maintaining a suspension by mixing to ensure good reaction and separation must not be at the expense of physical damage. The therapeutic properties of proteins produced from bioreactors using biotechnological processes such as recombinant DNA technology are dependent on their molecular conformation (or shape). Physical damage to the conformation by overaggressive mixing may otherwise compromise their therapeutic effectiveness.

What Else Is Interesting?

Recombinant DNA technology is a biotechnological process used to produce many useful and valuable products with medical, healthcare, agricultural, and veterinary applications. It involves extracting genes from the cells of a living organism and transferring them to the living cells of another organism. The genes transferred contain the genetic code for the expression of a required biochemical product, such as a protein with therapeutic properties, or provide resistance to an antibiotic or other substance. Recombinant human insulin, used for the treatment of diabetes, has virtually replaced insulin once produced from pigs and cows, and is produced by bacteria containing human genes. Human growth hormones, hepatitis B vaccine, blood-clotting agents known as Factor VIII, anti-cancer drugs, and vaccines against scours, which is toxic diarrhoea in pigs, are all produced by recombinant DNA technology. Crops developed through recombinant DNA technology which are resistant to herbicides, insects, low moisture, and other environmental conditions include rice, maize, canola, and cotton. The technology was pioneered by Stanley Cohen from the Stanford School of Medicine and Herbert Boyer from the California School of Medicine, San Francisco, in 1973.

Problem 7.6: Power for Mixing Scale-Up

A vessel is to be designed to contain a liquid slurry of suspended particulates using a mechanical stirrer. To ensure that the particulates remain
in suspension, scale tests are carried out under fully turbulent conditions
in a tank with a diameter of 0.6 m. The test tank has baffles and a flat-blade
impeller for which the tank diameter-to-impeller ratio is 3. The suspension is
considered to be successful for a mixing speed of 3 Hz for which the power
consumption is 0.18 kW and Reynolds number of 145,000. Following the success of the test mixing, determine the rotor speed that would ensure that the
same level of mixing is achieved at full scale if a geometrically similar vessel
is to be scaled-up by a linear scale of 6.

Solution

The slurry mixture consists of a viscous liquid consisting of a suspension of
solid particles. Maintaining suspensions is important in processes in which
sedimentation may be a problem or in which liquid–solid mass transfer is
important. Achieving off-the-bottom suspension aids mass transfer since
once the particles are in suspension there is intimate liquid contact. The mixing power can be expressed as

$$P_o = N_p \rho N^3 D^5 \tag{7.12}$$

For the test vessel which has an impeller diameter of 0.2 m,

$$N_p \rho = \frac{P_o}{N^3 D^5} = \frac{0.18}{3^3 \times 0.2^5} = 20.83\,\text{kWs}^3\text{m}^{-5} \tag{7.13}$$

The mixing power equation is therefore

$$P_o = 20.83\,N^3 D^5 \tag{7.14}$$

The impeller tip speed at the pilot scale is

$$U_1 = \pi N_1 D_1 = \pi \times 3 \times 0.2 = 1.88\,\text{ms}^{-1} \tag{7.15}$$

The impeller tip speed at the plant scale is required to have the same velocity.
Since the vessel is being scaled-up by a linear scale of 6, this corresponds to
an impeller diameter of 1.2 m. The speed of rotation of the rotor is therefore

$$N_2 = \frac{1.88}{\pi \times 1.2} = 0.5\,\text{Hz} \tag{7.16}$$

The scaled-up vessel therefore has a tank diameter of 3.6 m, impeller diameter of 1.2 m, and rotational speed of 0.5 Hz.

What Should We Look Out For?

Scale-up is the translation of a process design from the laboratory or experimental scale to the larger pilot-plant, commercial, or industrial scale and is an important part of commercializing a process in which it is accepted that theoretical design cannot always be used alone to achieve this. Dimensionless numbers or groups are a useful way of scaling-up a process since certain heat, mass, and momentum transfer phenomena are independent of scale. In mixing processes, it may be necessary to ensure that the power-to-volume ratio remains the same between the lab scale and full scale to ensure homogeneous mixing characteristics. The scaled-up equipment is also assumed to have geometric similarity. The testing of a small-scale process is therefore quick, cost-effective, and reliable such that the experimental information gained can be used on a larger scale. The Reynolds number at which the plant scale will operate is found from Equation 7.10. For constant physical properties of the liquid and assuming no change in operating conditions such as pressure and temperature, then

$$\frac{Re_2}{Re_1} = \frac{N_2 D_2^2}{N_1 D_1^2} \tag{7.17}$$

The new Reynolds number is therefore

$$Re_2 = Re_1 \frac{N_2 D_2^2}{N_1 D_1^2} = 14,500 \times \frac{0.5 \times 1.2^2}{3 \times 0.2^2} = 870,000 \tag{7.18}$$

The high Reynolds number will ensure that the particles are kept in suspension. However, if the solid particles have the propensity to settle at low impeller power inputs, or when the impeller stops either deliberately or unintentionally, for example, due to power failure, a solid sedimentary layer may form at the bottom of the tank. It may be impossible to restart the impeller without causing significant damage to the motor. When this is likely, either a gearbox or an alternative means should be used to suspend the sedimentary layer.

What Else Is Interesting?

In the scale-up of mixing vessels, the power number is related not only to the modified form of the Reynolds number but also to the Froude number presented in Problem 7.4. The performance of mixers in the food industry is typically expressed in terms of fluid velocity, total pumping capacity of the impeller, and total flow in the tank, or in terms of blending time or some readily evaluated solid-suspension criterion. If the application is relatively simple, a complete and detailed examination of the complicated fluid mechanics in a mixing tank is not usually necessary. If the process is complex, which is often the case, then a full analysis of fluid shear rates and stresses, two-phase mass transfer, turbulence, and microscale shear rate and blending will be required. This is important as there may be considerable cost implications if the qualitative and quantitative aspects of the complex mixing phenomena involved in mass transfer are not evaluated correctly.

Problem 7.7: Gas Bubbles in Mixing

A gas is dispersed within a liquid at a depth of 2 m in the form of small spherical bubbles with a uniform diameter of 2 mm. Determine the pressure within the bubble at this depth and the increase in radius at the point of reaching the surface. The surface tension is 0.073 Nm⁻¹.

Solution

Gases are usually dispersed within liquids as bubbles for the purpose of mass transfer, such as bubbling air in a fish tank to transfer oxygen. Gases can also be dissolved in liquids and released when the pressure of the liquid falls below the vapour pressure, as in the process of boiling. Carbon dioxide, which is a by-product of the brewing industry, is most commonly used in drinks as a preservative against microbial contamination and is very soluble in water in which one litre of water at 0°C can dissolve 1.8 litres of carbon dioxide at atmospheric pressure. When opening a bottle the pressure above the drink is released and provides the driving force for the carbon dioxide to come out of solution. Surface tension dictates that the pressure inside a bubble is inversely proportional to its radius. The pressure is also dependent on the pressure at the surface of the liquid and the hydrostatic depth:

$$p = p_s + \rho g h + \frac{2\sigma}{r} \tag{7.19}$$

The absolute pressure within the bubble with a diameter of 2 mm at a depth of 2 m is therefore

$$p = 101,300 + 1000 \times 9.81 \times 2 + \frac{2 \times 0.073}{0.002} = 120,993 \ \text{Nm}^{-2} \tag{7.20}$$

In comparison to the external pressure, the contribution from the surface tension is only 73 Nm⁻². As the bubbles rise to the surface, the hydrostatic pressure decreases such that there is an expansion of the gas. If the pressure and volume of the gas in the bubble are related by the simple relationship

$$pV = c \tag{7.21}$$

then the increase in volume for a spherical bubble is

$$p_1 r_1^3 = p_2 r_2^3 \tag{7.22}$$

The pressure in the bubble at the surface can therefore be found by solving the cubic equation

$$120,993 \times 0.002^3 = \left(101,300 + \frac{2 \times 0.073}{r_2} \right) \times r_2^3 \tag{7.23}$$

The cubic can be solved either analytically or by trial and error. If the radius is assumed to be very small, then the problem can be more readily solved. In this case, the radius is 0.00212 m.

What Should We Look Out For?

The pressure within a bubble is largely dominated by pressure at the surface of the liquid and to a certain extent the hydrostatic depth. However, an infinitely small bubble has an infinitely high internal pressure and will therefore not form spontaneously. Bubbles form by nucleation providing there is sufficient surface energy.

A change in the pressure on the surface of the liquid can influence the pressure within the bubble and therefore the volume. This forms the basis of the movement of the well-known Cartesian diver in which a small tube open at one end is immersed in a liquid contained within a flexible vessel such as a plastic bottle. A small amount of the liquid enters the tube that is able to allow the diver to be buoyant within the liquid. The diving effect occurs when pressure of the liquid is increased by squeezing the bottle, thereby compressing and reducing the volume of air in the tube, reducing the buoyancy and resulting in the descent of the sinker, until the pressure is released, whereupon the air expands and the diver rises again.

The increase in size of a bubble due to the decrease in hydrostatic depth is very small. However, the apparent rapid increase in bubble size of bubbles of carbon dioxide bubbles in carbonated drinks is due to the rapid rate of diffusion of carbon dioxide into the bubbles. Less clear is the influence of viscosity and surface tension on the bubbles, which can be seen in different carbonated drinks such as champagne, sparkling wines, beers, and lagers.

What Else Is Interesting?

The undesirable phenomenon of gushing occurs when the bottle or can of carbonated drink is opened rapidly, discharging its contents. This is caused by bubbles nucleating too rapidly. Shaking or tapping the bottle on a hard surface can stimulate this. It can also be caused by poor surface finish or a poor water quality containing particles.

The size of the bubbles and the rate at which they form depend on both the level of carbonation and the drink's ingredients. Mineral water contains little dissolved gas and tends to have large bubbles. Tonic water, on the other hand, is highly carbonated and produces a lot of very small bubbles. Tonic water therefore goes flat quickly. Champagne has a high level of carbonation and produces small bubbles that are released steadily over a long period of time. Sugar has an influence on both the surface tension and viscosity of the drink. These, in turn, affect the size and rate of formation of the bubbles. Alcohol also lowers the surface tension when the drink is used as a mixer.

Problem 7.8: Foams

Explain the phenomenon of foams and their formation in liquid mixing processes.

Solution

A foam is an aggregation of gas or vapour bubbles that remain as a stable suspension. The liquid is the continuous phase and exists as bubble films. Foaming is often a problem such as with fermentation processes that form on the surface of the broth. The existence of foam can be controlled using mechanical devices such as rotating blades or by the addition of chemicals such as glycol, which interrupt the stability of the foam.

Unlike the bubbles in a soft drink that burst at the surface, beers have a foamy head. The bubbles in some beers are stable and prevented from bursting when they reach the surface. The biggest bubbles tend to grow bigger at the expense of smaller bubbles because this allows them to lower their internal pressure and thus minimises the total surface area of the foam. Surface active agents can stabilise a bubble by forming a condensed phase of surfactant molecules packed round the bubble. Surfactants in beer include glycoproteins, which form stable surface layers by cross-linking and hydrogen bonding. Chemicals derived from the hops that make beer bitter are also responsible for making bubbles stable. Certain metal ions, including iron, are also stabilisers. Fats, however, reduce the stability of foams. Lipstick and oil from snacks are causes for the loss of head, which is why glasses must be washed scrupulously before reuse.

As bubbles rise to the under surface of the head, they push the existing foam upwards and the bubbles at the top of the head are squeezed together. Energetically, it is more favourable for them to share bubble walls. The bubbles therefore become more like pentagonal dodecahedrons than spheres. At the same time, liquid drains out of the rising foam concentrating the surfactants in the membranes and further stabilises the bubbles. The membranes are, however, permeable to gas, and the smaller bubbles, which have a higher internal pressure, diffuse gas into the larger bubbles. Nitrogen diffuses more slowly than carbon dioxide and so it provides a more stable foam.

Bottled beers tend to be particularly fizzy as they rely on bubble nucleation to generate a head. They therefore require plenty of dissolved gas. Draught beer, on the other hand, relies on a dispensing tap that contains fine holes. The turbulent flow during pouring creates bubbles by cavitation. These are then dispersed into the beer.

Cans of draught beer also rely on the mechanical dispersion of fine bubbles to produce a head. At the bottom of the can there is a flexible plastic capsule or *widget* within which a very small hole is pierced. During canning, the widget is filled with nitrogen and the beer added on top leaving a larger than usual space in the top of the can. The nitrogen remains inside the capsule because of its low solubility and the pressure inside the sealed can although some beer may actually enter through the hole. When the can is opened the release of pressure causes the gas to expand, passing out through the hole as a stream of tiny bubbles. These bubbles grow as they rise through the liquid and produce a head in the space at the top of the can. Unlike a bottled beer, these beers should be poured immediately.

Carbonation, which is critical in soft drink manufacture, involves the dissolution of carbon dioxide in aqueous media under pressure. This can be achieved either by dispersing the gas into liquids using conventional bubble columns or Venturi absorbers, or by dispersing liquid into a continuous gas as a spray or a film, using spray columns, packed columns, and wetted wall columns. In certain processes, such as whipping or whisking cream, or beating eggs, air may have to be included in a rheologically complex liquid phase. This can be achieved by the use of a combination of bubble diffusers and turbulence promoters, or by using vessel-type mixers wherein air is dispersed by whisks that are normally pear-shaped or cylindrical assemblies of tough stainless steel wires, similar to domestic blenders. The whisks rotate on shafts, which gyrate inside the vessel.

Air can also get inadvertently entrained into food materials during processing. This is a particular problem encountered while processing highly viscous materials such as ketchup. Air is undesirable in such products and once entrained, may be exceedingly slow to disengage. Product homogenisation by deaeration is also a mixing problem, and this can be achieved by using a slow-speed, upward pumping impeller, which facilitates air disengagement.

The purpose of homogenisation is to attain a more even particle size or a more homogeneous blend of materials. Applied on whole milk, homogenisation is intended to reduce the size of fat globules and to prevent skimming of fat. The liquid (whole milk) is pressed under high pressure (2–3 MNm^{-2}) through a small orifice.

What Should We Look Out For?

A froth is different from a foam and is a dispersion of a high fraction of gas or vapour in a liquid that readily separates at the surface of the liquid as bubbles disintegrate without forming a foam. Froth flotation is a process used in mineral dressing, which is the preparation of minerals and ores for extraction, and is used to separate ore and gangue (undesired and largely nonmetallic minerals of little commercial value) using small bubbles of air or another gas generated inside a tank containing particles of ore and gangue. Water is treated with a water-active agent (detergent), which allows either but not both materials to adhere to the bubbles. The bubbles rise and are skimmed from the surface.

What Else Is Interesting?

Where a foam forms which is considered to be undesirable, a foam breaker may be used. This is a mechanical device used to break up the formation of foam within a vessel. Foam breakers are sometimes used in fermentation vessels in which foaming may occur and otherwise overflow. They usually consist of high-speed spinning discs and are positioned in a region above a potentially foaming liquid. Where contamination of the liquid is not an issue, liquid foam breaking agents, such as glycol, can be used.

Problem 7.9: Mixing in the Food Industry

Describe the significance of mixing in the food industry.

Solution

Unlike the conventional mixing of liquids that are robust in a chemical or physical nature, over a wide range of operating conditions of temperature, pressure, and mixing shears, in which scale-up is based on power-to-volume ratios, the approach to mixing in the food industry requires a different approach. The intention of mixing foods is to attain a certain intimate mix of components which may be shear-sensitive, complex in composition, fragile, and with physical properties that are confined to a limited range of temperatures, such as proteins, fats, and carbohydrates as well as microcomponents such as vitamins. The outcome of mixing is often a function of the process and mixing history. For example, in making a cake or a batter, a different product results when the ingredients are mixed in a different order.

To achieve the required form of mixing, a wide variety of food mixers are available with an equally wide-ranging capability. Some are developed for specific applications, such as emulsification or solid dispersion into liquids, while others tend to be multi-purpose, and are used when a given mixer is required to perform a variety of mixing tasks. The design procedures employed are seldom based soundly on theory but instead, the success of a design for a given application is often the result of trying out innovative ideas based on past experience. Transferring information from one application to another is difficult.

The shear stress generated by an agitator is very important in a food system. Processes for making fine dispersions and emulsions require high shear, while others, such as mixing nuts and chocolates or whole fruit into yoghurt, require low shear. Many foods contain particles of different sizes, some of which are fragile. These particles may have to be dispersed into viscous liquids such as sauces. Agitation of such systems to produce uniform product composition is therefore very difficult. The segregation of particles often occurs when blended products are discharged from mixing vessels. Often, there are significantly higher coefficients of variation between packages than between samples taken from the mixer.

The mixing of food materials can often involve the addition of high value components, such as flavours in very small quantities into bulk materials. This process is particularly difficult when mixing into dry solids. It is also difficult to monitor and control the extent of mixing in foods, since the endpoint of agitation occurs some time after agitation has ceased. This is typical of processes involving gas inclusion, crystallisation, and texturisation.

Regardless of the shape or size of the mixer, its design features must ensure good hygiene, as well as provide for in-place cleaning, and, if necessary, sterile operation. Mixers used in the food-processing industries may be conveniently classified into categories according to the materials that are to be mixed. These include the dissolving and dispersing of liquids, blending of particulate material, and mixing of solids and liquids to form doughs, batters, and pastes.

In practice, solids, liquids, or gases may need to be dispersed and/or dissolved in liquids. Mixing is relatively simple if the particles are readily soluble. The dispersion of dense or buoyant particulates can pose considerable problems, however. Typical applications include reconstituting powdered milk, adding vitamins to liquid products, blending sugar with liquids, blending starch slurry for salad dressing, and adding stabilisers.

What Should We Look Out For?

Mixing devices used in the food industry that are currently available are generally effective for their intended applications. However, it is difficult to understand the mechanisms underpinning the operation of these devices, partly because of the complex motion that they impart but largely due to the complex properties of food systems that can themselves vary during mixing. As a result, developments in mathematical modelling of food mixing processes are virtually nonexistent. Established procedures for process design and scale-up are elusive. There appears to be much value in gaining an understanding of mixing mechanisms so that modelling techniques may be applied to the rational design of mixers.

What Else Is Interesting?

Emulsification of foods is one of the most complicated unit operations since the nature of the final product varies greatly depending on the method of preparation. Even the method of addition of components and the rate of addition can significantly affect the emulsion quality. Oil-in-water emulsions are widely used and may be produced in impeller-agitated vessels operating at high rotational speeds, colloidal mills, or high-pressure valve homogenisers. Continuous processing may be achieved by in-line mixers, which consist of a high-speed rotor inside a casing, into which the components are pumped and subjected to high shear. Emulsions are formed under extremely high specific power, typically around 10^{12} Wm^{-3}.

Problem 7.10: Power for Sparging

A stirred vessel has a volume of 600 litres and diameter of 25 cm. Air is sparged into the vessel at a rate of 100 litres per minute below the turbine in order to aerate the medium. Determine the power requirement due to air sparging into the vessel.

Solution

Aeration is the introduction and movement of air or oxygen at a low flow rate through a liquid medium such as a bioreactor or activated sludge process. Aeration is used to provide oxygen to microorganisms that are responsible for the biologically catalysed reactions. The oxygen is usually introduced through a sparger as small bubbles that have a high surface area. Aeration is therefore used to promote effective mass transfer of the oxygen to the liquid medium and therefore microorganisms. The power for sparging is the product of the airflow and pressure needed to be overcome due to the hydrostatic depth. Assuming that the bioreactor is not itself pressurised, the power is therefore

$$P_o = \dot{Q}\Delta p \tag{7.24}$$

The flow rate is related to the velocity and area, and the pressure is related to the hydrostatic depth. Since the volume of the bioreactor is the product of area and depth, the power is therefore

$$P_o = \frac{\dot{Q}}{a}\rho g V = \frac{0.1}{60} \times 1000 \times 9.81 \times 0.6 = 200 \text{ W} \tag{7.25}$$

That is, a power input of 200 W is required.

What Should We Look Out For?

As well as the rate of flow of air, the power input is used to overcome the hydrostatic depth, yet in this problem the depth is not apparent. This would suggest that it is independent of depth, but in practice, this is not the case. The sparger should be located at the bottom of the bioreactor to provide sufficient hold-up for aeration.

What Else Is Interesting?

There are various sparger designs, but all involve the release of air or a gas into a liquid medium such that small bubbles are created. Some consist of open pipes or tubes, while others consist of perforated tube walls. Figure 7.9

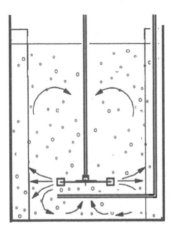

FIGURE 7.9
Bubble Dispersion within a Vessel

illustrates the circulation of bubbles of air from a sparger located below an impeller in a vessel with baffles.

Problem 7.11: Optimisation of Power Input in Stirred Tanks

A stirred tank bioreactor with a volume of 30 m³ is to be operated by sparging with oxygen gas with oxygen mass transfer coefficient ($k_l a$) of 0.054 s⁻¹. The impeller has a diameter of 1.12 m and a blade width of 20 cm. Determine the operating conditions in terms of the aeration rate ($\dot{Q}$) and impeller speed (N) that minimises the power input to the bioreactor. The power input through the impeller is given as

$$P_g = 0.1P\left(\frac{NV}{\dot{Q}}\right)^{0.25}\left(\frac{N^2 D^4}{gW_b V^{\frac{2}{3}}}\right)^{-0.2} \tag{7.26}$$

where P_g is the gassed power input, P is the ungassed power input, D is the impeller diameter, N is the impeller speed, $\dot{Q}$ is the aeration rate, g is the gravitational acceleration, W_b is the impeller blade width, and V is the bioreactor volume. The overall oxygen mass transfer coefficient is given by $k_L a = 0.4e$, where e is the fractional gas hold-up given by

$$e = 0.113\left(\frac{\dot{Q}N^2}{\sigma}\right)^{0.57} \tag{7.27}$$

where σ is the surface tension of the broth (0.05 Nm⁻¹). The impeller power number N_p is 5 for a Reynolds number greater than 5000. The viscosity of the broth is 0.02 Nsm⁻², and the density is 1000 kgm⁻³. The power (Watts) due to sparging is related to the rate of flow of oxygen (m³s⁻¹) and may be given by $P = 6440\dot{Q}$.

Solution

In this problem, the power is the sum of both sparging and stirring. In the first, the power can be determined from the fractional gas hold-up of the sparged air which is found from the overall mass transfer coefficient. The $k_L a$ is the volumetric mass transfer coefficient, which is seen as a measure of the transfer of a gas to a liquid in processes such as fermentation in which oxygen or air is sparged into a liquid containing microorganisms. It is the product of the mass transfer coefficient, k_L, and the interfacial area, a. It is difficult to measure the mass transfer coefficient separately, but it is readily measured from gas balances in combination with the interfacial area. In this case,

$$e = \frac{0.054}{0.4} = 0.135 \tag{7.28}$$

Therefore from Equation 7.27,

$$\dot{Q} = \frac{\sigma}{N^2}\left(\frac{e}{0.113}\right)^{\frac{1}{0.57}} = \frac{0.02}{N^2} \times \left(\frac{0.135}{113}\right)^{\frac{1}{0.57}} \tag{7.29}$$

This reduces to

$$\dot{Q} = 0.0683 N^{-2} \tag{7.30}$$

The total power is the sum of the sparging and stirrer power

$$P_T = 6440\dot{Q} + 0.1P\left(\frac{NV}{\dot{Q}}\right)^{0.25}\left(\frac{N^2D^4}{gW_bV^{\frac{2}{3}}}\right)^{-0.2}$$

$$= 6440 \times 0.0683 N^{-2} + 0.1 \times P \times \left(\frac{N \times 30}{0.0683 N^{-2}}\right)^{0.25}\left(\frac{N^2 \times 1.12^4}{9.81 \times 0.2 \times 30^{\frac{2}{3}}}\right)^{-0.2} \tag{7.31}$$

where the power for stirring is given by Equation 7.12. The value of N_p is given as 5 to give

$$P = 8812N^3 \tag{7.32}$$

The total power for sparging is therefore

$$P_T = 439.9N^{-2} + 6636N^{3.35} \tag{7.33}$$

This parabolic variation of stirrer speed with power can be readily differentiated to show that the minimum power is 2.35 kW with an aeration rate of 0.226 m^3s^{-1} as illustrated in Figure 7.10.

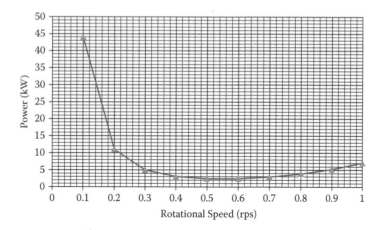

FIGURE 7.10
Variation of Pump with Shaft Speed

What Should We Look Out For?

Dispersing gases within a liquid in the form of bubbles is often done in order to transfer the gas or a component within the gas into the liquid. The intimate contact of a gas in the form of bubbles within a liquid is particularly sensitive to the design of the mixer that is able to create the necessary interfacial area and consequently permit the mass transfer. In specifying a mixer for this purpose, the description of the mechanical dispersion of the gas is usually required. This is not, however, a good substitute for the mass transfer or reaction rate.

The mixing system is designed to provide good dispersion of the bubbles and to avoid geysering effects resulting in poor mass transfer and surface disruption effects. Dispersion can be achieved in many ways. The use of radial-flow impellers, in which the gas is usually admitted below the rotating impeller that has sufficient power input and rotational speed, creates bubbles that are able to travel within the currents formed by the impeller blades. With sufficient impeller power, the currents, and hence bubbles, are driven out to the side of the vessel with a portion circulating down to the bottom of the tank.

The injection of the gas may be in the form of an open pipe or tube. A sparge ring is often used to create the intimate dispersion of bubbles, which consists of a circular shaped tube perforated with holes and is usually located beneath the impeller. The nature of the design and operation of the impeller determine the bubble size and interfacial area for mass transfer rather than the number and size of the holes. The size and proximity of the sparge ring is, however, important. In general, the ring diameter should be around eight-tenths of the impeller diameter so that the gas can enter the high-shear zone directly, which is located at the impeller tips where the major recirculation of the gas occurs. Where plugging or corrosion of the ring may be a problem, an open pipe may be preferentially used. (See also Problem 7.10.)

What Else Is Interesting?

A single turbine located at the bottom of a tank may not be sufficient for adequate dispersion and mass transfer where the bubbles may tend to coalesce. The use of multiple turbines makes it possible to maintain a small bubble size for relatively tall mixing vessels. The spacing of the turbines can be much closer than for other fluid-mixing applications, which may be typically one and a half to two impeller diameters.

An effective method of dispersion and mass transfer is to use a radial impeller at the bottom of the vessel with two or three axial-flow turbines located in the upper section of the vessel. These upper turbines are used for pumping the dissolved solute from the gas phase in the upper part of the tank. The more common method, however, is to use radial-flow impellers only. For any particular application, however, both methods should be considered, and the most practical one should be selected.

The upward flow of gas from a sparger virtually destroys the flow pattern created by axial-flow turbines and therefore does not provide the required flow pattern for dispersion. An exception to this is to use a draft-tube aerator in which the turbine is confined within the tube and has sufficient power to overcome the upflow of gas. A net downflow of gas bubbles to the bottom of the tube results, providing effective mass transfer. This is typically used in wastewater treatment.

One of the most common requirements for the dispersion of gases within liquids is to transfer a certain quantity of gas, or a particular component within a gas, per unit of time. The basic measurement in mass transfer is the mass transfer rate, which is usually expressed as the product of the volumetric coefficient and the average concentration driving force. There are many complexities required in determining the mass transfer driving force. Henry's law relates the partial pressure in the gas phase with the concentration in the liquid phase. For many gas–liquid combinations, however, Henry's law constants are not known. A decision must nevertheless be made about whether to use a gas concentration or a liquid concentration driving force. If a large part of the mass transfer occurs in the impeller region, the dissolved gas concentration may approach saturation such that any further increase in power may not improve mass transfer performance.

Problem 7.12: Scale-Up

A laboratory scale stirred vessel with a capacity of 12 litres is used to carry out tests on the mixing of a liquid. The vessel is cylindrical and agitated using three Rushton-type impellers on the same shaft which rotates at 600 rpm. The aspect ratio of the reactor has a height-to-diameter ratio of 3:1, and the impeller diameter is 30% of the vessel's diameter. It is intended to scale-up the mixing process, which is to be carried out in a 10,000 litre stirred vessel. Determine the agitator speed for a constant power-to-volume ratio, for a constant impeller tip speed, and for a constant Reynolds number.

Solution

The aspect ratio is the relative dimension of an item of process plant equipment such as the ratio of the height to width or diameter of a column or storage tank. In this case, the dimensions of the laboratory scale vessel are

$$V = \frac{\pi D^2}{4} H \qquad (7.34)$$

With the height being three times that of the diameter, and the total volume being 12 litres, the dimensions are therefore a diameter of 0.172 cm, height of 51.6 cm, and impeller diameter of 5.2 cm. In scaling-up the vessel for the same aspect ratio, the dimensions of the 10,000-litre reactor are therefore a diameter of 1.72 m, height of 5.16 m, and impeller diameter of 0.52 m.

For the agitator operating at the same power-to-volume ratio, the power consumption is given by Equation 7.12 from which

$$N_1^3 D_1^2 = N_2^3 D_2^2 \qquad (7.35)$$

Since the scaled-up volume is proportional to the impeller diameter, the scaled-up impeller speed is therefore

$$N_2 = \sqrt[3]{N_1^3 \frac{D_1^2}{D_2^2}} = \sqrt[3]{600^3 \frac{0.172^2}{1.72^2}} = 129 \text{ rpm} \qquad (7.36)$$

For the case of the scaled-up vessel to operate at the same impeller tip speed, the velocity at the tip is

$$U = \pi D N \qquad (7.37)$$

That is, the tip speed is proportional to DN. Considering both vessels, the agitator speed of the 10,000-litre vessel is 60 rpm.

For the case of the scale-up requiring the same Reynolds number (Equation 7.10), then where the density and viscosity are unchanged, the scale-up rotational speed is

$$N_2 = N_1 \left(\frac{D_1}{D_2} \right)^2 = 600 \times \left(\frac{0.052}{0.52} \right)^2 = 6 \text{ rpm} \qquad (7.38)$$

What Should We Look Out For?

In general, dimensionless numbers are useful in the scale-up of process equipment. It is, however, important to have a wider appreciation of the nature of the materials being mixed. For example, the shear sensitivity of biochemicals may determine the maximum shear stress that can be imposed by the tip of the rotating blades of the impeller, or the complex rheological properties of a polymer may determine the type of impeller and rotational speed.

What Else Is Interesting?

In the food industry, the performance and scale-up of mixing processes is dependent on the history of the mixing process, the order and rate at which ingredients are added. The process is also often time dependent in which the properties of the mixture change with time. Mixers are also typically expressed in terms of fluid velocity generated, total pumping capacity of the impeller, and total flow in the tank, or in terms of blending time or some readily evaluated solid-suspension criterion. If the application is relatively simple, a complete and detailed examination of the complicated fluid mechanics in a mixing tank is not usually necessary. If the process is complex, which it usually is, then a full analysis of fluid-shear rates and stresses, two-phase mass transfer, turbulence, and microscale shear rate and blending may be required. This is important as there may be considerable cost implications if the qualitative and quantitative aspects of the complex mixing phenomena involved in mass transfer are not evaluated correctly.

Further Problems

1. Explain the purpose of baffles used within vessels for the mixing of liquids.
2. With the use of examples, explain the criteria used for the successful mixing of multi-component mixtures.
3. In terms of flow patterns, explain the difference between the use of impellers and propellers for the mixing of liquids in vessels.
4. A 10-litre stirred test vessel is to be scaled-up to 1000 litres. The test vessel operates using an impeller at 180 rpm. The aspect ratio of the

vessel is 3:1, and the impeller diameter is 33% of the vessel's diameter. Determine the agitator speed for a constant power-to-volume ratio. *Answer*: 39 rpm

5. A stirred vessel with a volume of 100 litres and diameter of 40 cm contains an aqueous solution that requires aeration. If the aeration rate is 10 vvm (volumes of air per volume of vessel per minute), determine the power required for aeration. *Answer*: 1.3 kW

6. Determine the pressure in a spherical bubble of gas in water which has a diameter of 1 mm. The surface tension is 0.071 Nm^{-1}. *Answer*: 142 Nm^{-2}

8

Particle Flow

Introduction

The study of particle flow concerns the relative motion between a particle and the fluid within which it is suspended. The particle is often considered to be a solid having a small diameter but equally, it may be considered as a small liquid droplet in a gas or, conversely, a small gas bubble in a liquid. Particles may be the product emanating from a process waste stream, such as dirt, dust, smoke, soot, fumes, aerosols, mists, and sprays. There are many examples of fluids that transport or carry particles. Industrial examples include settling tanks used for clarifying dirty water, centrifuges, dust handling and collection in electro-static precipitators, cyclones, particle size analysis, sprays of liquids including spray dryers and oil burners, and hydraulic flocculation. Most engineering design calculations are based on the terminal velocity of a particle. This is the point in which a single particle falling through a fluid under the influence of gravity or some other force reaches the point where it is balanced by frictional resistance and where there is no acceleration.

In practice, particles rarely have a uniform shape and size. The particle size distribution is a classification of size based on their diameter or some other physical characteristic such as weight, which is usually expressed as the quantity of the particles whose size fall between two measureable characteristics such as the weight percentage of the mixture between two defined diameters. It can also be expressed as a cumulative quantity such as the total weight expressed as a percentage of the mixture above a certain size.

Problem 8.1: Stokes' Law

Determine the time it takes for a single spherical particle with a diameter of 100 μm and density of 2800 kgm^{-3} to settle in an oil with a density of 900 kgm^{-3} and viscosity of 0.08 Nsm^{-2}.

Solution

The terminal velocity of a particle is the velocity that has been attained when the external forces on the body, for example, due to gravity are balanced by the resistive drag forces of the surrounding fluid. Particles can be separated from liquids at their terminal velocity under the influence of gravity in settling tanks such as lagoons. Where the terminal velocity is slow under the influence of gravity alone and/or where a more rapid separation is required, external forces can be applied, such as centrifugal, electro-static, and magnetic forces. This results in an increase in the terminal velocity and, consequently, a reduction in the time for separation.

The general force equation on a particle suspended in a fluid involves external, buoyancy, and drag forces. Buoyancy force depends on the Archimedes' principle, which states that when a body is immersed in a fluid it experiences an upthrust equal to the weight of the fluid displaced. Drag force is determined from the resistance caused by the fluid. The terminal velocity is given as follows:

$$U_t = \frac{g\left(\rho_p - \rho\right)d_p^2}{18\mu} \tag{8.1}$$

Known as Stokes' law, it is named after the Irish physicist and mathematician noted for his contributions to fluid mechanics, Sir George Gabriel Stokes (1819–1903). In this case,

$$U_t = \frac{9.81 \times (2800 - 1000) \times (100 \times 10^{-6})^2}{18 \times 0.08} = 1.23 \times 10^{-4} \text{ ms}^{-1}$$

What Should We Look Out For?

The conditions for Stokes' law assume an infinite dilution in which the proximity of other particles that interfere with the movement can be neglected. There is also a minimum size of particle for the fluid viscosity to operate. Brownian motion is the small, irregular, and continuous movement of very small particles suspended within a fluid. Particles with a diameter of less than one micrometre (1 μm) have a random movement caused by collisions with other particles. This is named after the British botanist, Robert Brown (1773–1858), who first noticed the phenomenon while studying pollen particles.

Stokes' law is formulated from a balance of forces in which the drag force on sufficiently small particles is assumed to be inversely proportional to the particle Reynolds number. The acceleration of particles is therefore given by

$$\frac{dU}{dt} = g\left(1 - \frac{\rho}{\rho_p}\right) - \frac{18\mu U}{\rho_p d_p^2} \tag{8.2}$$

The particle accelerates from stationary and continues to accelerate and will according to this equation achieve terminal velocity after an infinite period of time. It is more convenient, however, to express the terminal velocity as some proportion of the terminal velocity, and particularly in the design and operation of industrial processes and equipment. For example, a 100 μm diameter particle, for example, may achieve 90% of its terminal velocity in about 300th of a second. Therefore, it can be seen that when designing equipment with the times involved measured in seconds, the time taken for particles to achieve their terminal velocity can sensibly be neglected.

There are cases where the time to reach terminal velocity is slow or intentionally not permitted such as the operation of equipment that uses pulsed flow. If the pulse is very short, then the particle may not actually achieve its terminal velocity since it will be accelerating and decelerating with the flow of the fluid. Many types of mineral processing equipment separate heavy ore particles from lighter particles by this means.

For small particles, the time taken to achieve 99% of its terminal velocity and the distance travelled can be readily found. It can also therefore be deduced that small, single particles get very close to their terminal velocities in a very short period of time and that the distance travelled by single particles is very small while achieving terminal velocity. For particle Reynolds numbers that exceed 0.4 but fall below 500, the single-particle terminal velocity is given by

$$U_t = \left(\frac{4}{225} \frac{(\rho_p - \rho)^2 g^2}{\rho \mu} \right)^{\frac{1}{3}} d_p \tag{8.3}$$

This is of more interest for cases where large particles are critical, such as with sprays. Newton's law exists at even higher particle Reynolds numbers of up to 200,000 where the terminal velocity is given by

$$U_t = \left(\frac{3.1(\rho_p - \rho) g d_p}{\rho} \right)^{\frac{1}{2}} \tag{8.4}$$

This is not generally of interest in most engineering applications and is a region where the particle drag coefficient is independent of the particle Reynolds number for which a value of 0.44 is often used.

The use of analytical expressions to determine terminal velocity requires trial-and-error procedures. This requires the particle Reynolds number to be determined after calculating the terminal settling velocity to determine the validity of the appropriate equation.

What Else Is Interesting?

There are many applications of Stokes' law. An example is the separation of particles in settling tanks. These are used by industries that use large quantities of water and other liquids which may become contaminated with solid particles, droplets of an immiscible liquid, or gas bubbles. These contaminants are required to be removed before the liquid can be recycled or discarded such as in the processing of sugar beets, where jets of water remove soil, which is then separated from the water into large settling tanks.

A spray dryer is a device that is another example of particle flow used to remove the moisture from a high moisture-containing fluid that contains a solid to be dried. The solid is often heat labile, such as milk, and is continuously atomized into small droplets within a large chamber. The chamber also has a continuous flow of warm, drying air or gas. Evaporation of the suspended droplets is rapid, and the dried product is quickly carried away with the current of air or gas and separated, usually in a cyclone. Rapid drying in this way is suitable for materials that may be heat sensitive such as certain biological and food products. Atomization is achieved using a device that produces very small droplets of a liquid. Such small droplets can be produced by forcing a liquid through a very small aperture under high pressure or by contacting the liquid with a high-speed rotating plate or disc. In atomizing the liquid into small droplets within the chamber of the dryer, the drying effect is sufficiently rapid such that the moisture is removed, leaving dried particles before they can otherwise reach and adhere to the walls of the dryer. The dried particles can then be collected from the bottom of the dryer. Depending on the design and operation of the atomiser, as well as the properties of the liquid (density, viscosity, surface tension, and solids content), a range of droplets can be formed. Empirical relations are available to estimate the size of the particle. Visual measurement techniques are also used. In practice, the operation of the spray dryer is based on experience to balance the mass and energy requirements for drying.

Problem 8.2: Particle Settling in Lagoons

A lagoon is used to continuously separate particles of waste ore in a wash water. The density of the ore is 2800 kgm^{-3}, and the particles are suspended in the flow of the wash water at a rate of 360 m^{3}h^{-1}. The lagoon is rectangular, being 10 m in width and 50 m in length, with a depth of 2 m. Determine the size of the particle that can be retained by the lagoon if the particles are assumed to be spherical and the maximum particle size obeys Stokes' law for a particle Reynolds number that is below a value of 0.4. The density of water is 1000 kgm^{-3} and the viscosity is 0.001 Nsm^{-2}.

Solution

A lagoon is a wide and shallow body of water used for settling and separating fine particles from process liquids under the influence of gravity. It can also be used for collecting process water and other collected water prior to discharge. Settling or sedimentation tanks are also used for a similar purpose. The design is based on a steady flow entering and leaving in which the bulk flow is along the length and uniform at all points in the lagoon with Stokes' law applied for the vertical velocity component. The capacity of the lagoon is designed such that any particle touching the base of the tank will be retained by the tank, and conversely, any other particle will be swept out with the effluent. The settling zone is the largest region in a lagoon used to separate solid particles from a liquid. It is the largest portion of the tank and provides a calm area in which the suspended particles are able to settle. Below is the sludge zone in which the particles accumulate.

By considering a single particle in a flow of water, the particle is expected to be retained by the lagoon, of width W, depth H, and length L, and thus reach the bottom under the influence of gravity before reaching the far end. An allowance is usually made for the depth of the scrapper mechanism and allowances at either end are made for water entering and leaving the system. The design is based on a number of assumptions and simplifications. The flow is assumed to be steady on both entering and leaving the lagoon for which the liquid velocity is uniform at all points. It also assumes that any particle touching the base of the lagoon will be retained, and conversely, any other particle will be swept out with the effluent. By ignoring the acceleration of the settling particle to reach its terminal velocity of the water, which is negligible, the average velocity of the water along the lagoon, and hence particle velocity, is based on the flow rate and cross-sectional area:

$$U_x = \frac{\dot{Q}}{WH} = \frac{360/3600}{10 \times 2} = 0.005 \text{ ms}^{-1} \tag{8.5}$$

The entrapment of a particle therefore requires that the particle must settle within a time period of

$$t = \frac{L}{U_x} = \frac{50}{0.005} = 10,000 \text{ s} \tag{8.6}$$

This is known as the contact or space time. The limiting or critical size of particle, d_c, is the time which will have a net velocity giving a trajectory along the diagonal of the tank. All particles of d_c or greater will therefore be retained by the tank. In this case, the terminal velocity of the particle can therefore not exceed

$$U_t = \frac{H}{t} = \frac{2}{10,000} = 2 \times 10^{-4} \text{ s} \tag{8.7}$$

Note that a small particle size, d_p, will have a lower terminal velocity than the critical particle size, d_c. Assuming a density and viscosity of the water of 1000 kgm⁻³ and 0.001 Nsm⁻², respectively, the critical diameter of the particle is therefore found by rearranging Stokes' law (Equation 8.1):

$$d_p = \sqrt{\frac{18\mu U_t}{g(\rho_p - \rho)}} = \sqrt{\frac{18 \times 0.001 \times 2 \times 10^{-4}}{9.81 \times (2800 - 1000)}} = 1.42 \times 10^{-5} \text{ m} \tag{8.8}$$

or 14.2 µm. If the validity of Stokes' law is based on a particle Reynolds number, Re_p, of up to a value of 0.4, then the terminal velocity of a particle is therefore

$$Re = \frac{\mu}{\rho_p U_t d_p} = 0.4 \tag{8.9}$$

That is,

$$U_t = \frac{g(\rho_p - \rho)d_p^2}{18\mu} = \frac{\mu}{0.4\rho_p d_p} \tag{8.10}$$

Therefore, the particle diameter cannot exceed

$$d_p = \sqrt[3]{\frac{18\mu^2}{0.4\rho_p(\rho_p - \rho)g}} = \sqrt[3]{\frac{18 \times 0.001^2}{0.4 \times 2800 \times (2800 - 1000) \times 9.81}} = 9.69 \times 10^{-5} \text{ m} \tag{8.11}$$

that is, 97 µm.

What Should We Look Out For?

The lagoon will accumulate with particles that settle as a bed of growing depth. The flow area will therefore become increasingly restricted, and the superficial velocity will rise to the point that particles will no longer be able to be retained and will consequently wash out. The lagoon operation is therefore required to be halted and the sludge or bed dug and scraped out.

Settling lagoon and tanks, also known as sedimentation tanks or clarifiers, are used in water supply and wastewater treatment systems. In drinking-water treatment, coagulants are added to the water prior to sedimentation to facilitate the settling process. This is then followed by various types of filtration and other treatment steps. They are used to provide some degree of purification. In sewage treatment plants, the water usually first enters through screens and grit chambers to remove large objects and coarse solids. The layer of accumulated solids at the bottom is known as sludge. The primary sedimentation is then followed by secondary treatment and consists

FIGURE 8.1
Silt Trap (Photo from Emily Schaschke.)

of trickling filters or activated sludge systems used to increase the level of purification further.

What Else Is Interesting?

Silt traps are used to separate out and capture silt, grit, and organic matter from storm water, streams, and other watercourses before flowing into drains, ponds, and lakes where their presence would have a detrimental effect. They therefore contribute to improving water quality as well as preventing material loss through erosion. Various designs are commonly used and range from excavated partitioned sumps with overflows to temporary permeable retaining barriers typically used at construction sites (Figure 8.1).

Problem 8.3: Particle Acceleration

A spherical particle of calcium carbonate with a diameter of 100 μm and of density 2800 kgm⁻³ settles freely in water. Determine the time it takes for the particle to reach 90% of the terminal velocity. The density and viscosity of water may be taken as 1000 kgm⁻³ and 0.001 Nsm⁻².

Solution

The acceleration of a particle is derived from a force balance on a particle and is related to the buoyancy effect and drag resistance expressed as

$$\frac{dU}{dt} = g\left(1 - \frac{\rho}{\rho_s}\right) - \frac{18\mu U}{\rho_s d_p^2} \tag{8.12}$$

The terminal velocity is given by Stokes' law (Equation 8.1), which occurs at an infinite period of time:

$$U_t = \frac{g(\rho_s - \rho)d^2}{18\mu} = \frac{9.81 \times (2800 - 1000) \times (100 \times 10^{-6})^2}{18 \times 0.001} = 9.81 \times 10^{-3} \text{ ms}^{-1} \quad (8.13)$$

Since the time to attain this velocity takes an infinite period of time, 90% of this value therefore corresponds to a velocity of 0.00883 ms^{-1}. Integrating Equation 8.12 therefore gives a time to reach this velocity of

$$t = \frac{1}{18\mu} \ln \left(\frac{g\left(1 - \dfrac{\rho}{\rho_s}\right) - \dfrac{18\mu U}{\rho_s d_p^2}}{g\left(1 - \dfrac{\rho}{\rho_s}\right)} \right) \quad (8.14)$$

$$\frac{1}{\dfrac{18 \times 0.001}{2800 \times (100 \times 10^{-6})^2}} \ln \left(\frac{9.81 \times \left(1 - \dfrac{1000}{2800}\right) - \dfrac{18 \times 0.001 \times 0.00883}{2800 \times (100 \times 10^{-6})^2}}{9.81 \times \left(1 - \dfrac{1000}{2800}\right)} \right) = 3.68 \times 10^{-3} \text{ s}$$

That is less than four-thousandths of a second.

What Should We Look Out For?

The particle will theoretically never reach its terminal velocity since this is approached asymptotically. It is therefore usual to express the time to reach a velocity as some fraction or percentage of the terminal velocity determined from Stokes' law. This sets the change in velocity of the particle with time to zero:

$$\frac{dU}{dt} = 0 \quad (8.15)$$

What Else Is Interesting?

A skydiver jumping from an aeroplane will accelerate under the influence of gravity to the point that the resistance of the air eventually balances the skydiver's weight. The terminal velocity reached by the skydiver typically ranges from around 50 ms^{-1} to nearly 80 ms^{-1} depending on the physical size and orientation of the skydiver. A horizontal position with limbs spread wide has a lower terminal velocity than a head-first position with a compact shape. When Austrian skydiver Felix Baumgartner jumped from a helium balloon at an estimated altitude of 39 km, on October 14, 2012, he reached an estimated terminal velocity of 377 ms⁻¹ or Mach 1.25 due to the very low resistance of the atmosphere in the stratosphere. He was the first person to break the sound barrier without mechanical assistance.

Problem 8.4: Particle Separation by Elutriation

Particles of ore suspended in water are to be separated by elutriation. A small test rig comprising a vertically mounted cylinder with an internal diameter of 5 cm is to be used to determine the effectiveness of separation in which a flow of particle-bearing water is continuously fed. For a rate of flow to the elutriator of 300 mL min⁻¹, determine the effectiveness of the elutriator in terms of particle sizes that can be separated and retained.

Solution

Elutriation is the process of separating suspended particles in a liquid by the upward flow of liquid such that the smaller particles with insufficient buoyancy are washed out or elutriated. It is typically used for separating particles into different size fractions. The particles are assumed to obey Stokes' law (Equation 8.1) for which the terminal velocity of the descending particles is equal to the upward velocity of the water fed to the separation device. The Reynolds number is less than 2000 and is expressed in terms of the flow rate to confirm that the upward flow is laminar:

$$Re = \frac{4\rho\dot{Q}}{\pi d\mu} = \frac{4 \times 1000 \times 300 \times 10^{-6}/60}{\pi \times 0.05 \times 0.001} = 127 \tag{8.16}$$

Within the circular cross section of the cylindrical elutriator, the maximum velocity of the upward flow occurs along the centreline, which for laminar flow corresponds to twice the average velocity. This is

$$U_{max} = \frac{2\dot{Q}}{a} = \frac{8\dot{Q}}{\pi d^2} = \frac{8 \times 300 \times 10^{-6}/60}{\pi \times 0.05^2} = 5.09 \times 10^{-3} \, ms^{-1} \tag{8.17}$$

This corresponds to the terminal velocity of the particles, which based on Stokes' law corresponds to a critical size of

$$d_p = \sqrt{\frac{18\mu U_t}{g(\rho_p - \rho)}} = \sqrt{\frac{18 \times 0.001 \times 5.09 \times 10^{-3}}{9.81 \times (1800 - 1000)}} = 1.08 \times 10^{-4} \, m \tag{8.18}$$

That is a diameter of 108 μm. This is the largest size of particle that can be elutriated corresponding to the highest terminal velocity when it is located at the centre of the elutriation tube. That is, it is furthest from the walls of the tube.

What Should We Look Out For?

All sizes below this determined size will be elutriated, although by assuming laminar flow, slightly smaller particles away from the centreline and closer to the wall where the local velocity is less may be elutriated. The velocity of the liquid in the cross section with laminar flow has a parabolic variation with a radius given by

$$U = \frac{1}{4\mu} \frac{\Delta p}{L} \left(R^2 - r^2 \right) \qquad (8.19)$$

The rate of laminar flow of a Newtonian fluid is expressed by the Hagen-Poiseuille equation:

$$\dot{Q} = \frac{\pi}{8\mu} \frac{\Delta p}{L} R^4 \qquad (8.20)$$

Eliminating the pressure drop per unit length of the tube gives

$$U = \frac{2\dot{Q}}{\pi} \left(\frac{R^2 - r^2}{R^4} \right) \qquad (8.21)$$

For elutriation, this is equal to the terminal velocity and corresponds to the particle sizes shown in Table 8.1.

What Else Is Interesting?

An air elutriator is another type of device that is used to separate particles into different sizes. It consists of a vertical tube up in which air flows. The particles for separation are carried at their terminal velocity in the upward flow. Particles below a certain size are collected as the overflow. Once all the particles of a certain size have been separated, the flow rate can be increased and another size fraction collected or by using tubes of differing cross sections arranged in series.

TABLE 8.1

Elutriator Particle Size Distribution

Radius from centreline (cm)	0	0.005	0.010	0.015	0.020	0.025
Terminal velocity ($\times 10^{-5}$ ms^{-1})	509	489	428	326	183	0
Max particle size retained (μm)	108	106	99	86	65	0

Problem 8.5: Anomalies in Particle Settling

Provide practical examples of anomalies where particles do not appear to obey Stokes' law.

Solution

Stokes' law assumes a particle or body is able to freely descend with an infinite dilution of the fluid within which it is suspended. This means that it is not affected by nearby objects as other particles or static surfaces. Providing that the other requirements of Stokes' law are also met, such as a limitation on the Reynolds number, a particle that has reached its terminal velocity can be timed to descend over a fixed distance from which the viscosity of the liquid can be determined, providing the liquid is sufficiently clear to permit the visibility of the movement of the particle. An improvement on the design is to use a sinker within a transparent tube and to time the descent between two points a fixed distance apart. The tube may be open or closed, but either way, as the sinker descends there is a displacement of liquid up around the sinker through the annular gap formed between the sinker and the tube. With the sinker descending with laminar flow, the fluid in contact with the wall of the sinker is assumed to be stationary, and equally, the fluid in contact with the moving sinker has the same velocity as the sinker. Both are called the *no-slip condition*. The result is that as the sinker descends, it draws the liquid nearby downward, yet for the sinker within the enclosed tube, there is a flow upwards through the annulus at a rate equal to the displacement volume of the sinker descending downward. At some point across the gap between the sinker and the wall, there is a flow upward as shown in Figure 8.2. The fast descent of large particles which provides a net upflow of displaced fluid carrying smaller particles upward is the cause of sedimentary settling in which the large particles form a layer at the bottom and the finer particles eventually settle on top. This phenomenon occurs in many particle settling

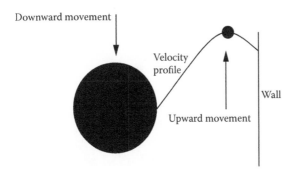

FIGURE 8.2
Velocity Profile for a Large Particle Descending and Small Particle Ascending

applications such as in the food industry in which canned foods that feature particles such as meat of various sizes in a sauce form settled layers.

What Should We Look Out For?

The velocity profile shown in Figure 8.3 is found in closed falling body viscometers. The principle of operation relies on gravity to provide the external force in which an object is allowed to descend freely. The fluid surrounding the body provides frictional resistance in which the forces balance, whereupon the body attains a terminal and constant velocity. This principle forms the basis of all falling body viscometers in which the body is permitted to descend freely through a fluid undergoing testing. The object may typically be in the form of a sphere, a cylinder, or a needle. Cylinders and needles are usually allowed to descend axisymmetrically. The velocity profile can be determined for fully developed laminar flow down a closed vertical tube with a terminal velocity based on Navier-Stokes equations. For a closed system, there is a displacement of liquid up the gap as the sinker descends. The velocity profile between the sinker and the tube wall can be deduced to be

FIGURE 8.3
Paraglider with Constant Velocity (Photo from C.J. Schaschke.)

$$U = \frac{1}{2\mu}\frac{dp}{dz}\left(\frac{R^2 - r^2}{2} - r_{max}^2 \ln\frac{r}{R}\right) \tag{8.22}$$

What Else Is Interesting?

The Navier-Stokes equations are a set of mathematical expressions used to study the motion of fluids. They are expressed in terms of velocity gradients for a Newtonian fluid with constant density and gradient. Using Cartesian or rectangular coordinates (x, y), the equations represent the inertia of body force, pressure, and viscous terms. The equations can also be written using cylindrical and spherical coordinates. The solutions to the equations are called *velocity fields* or *flow fields*. The equations were developed by Claude-Louis Navier (1785–1836) in 1822 and developed further by George Stokes (1819–1903) and find many applications including the study of the flow of fluids in pipes and over surfaces.

Problem 8.6: Fluidized Bed

A fluidized bed consists of spherical catalyst particles of uniform size with a diameter of 1.5 mm and a density of 3000 kgm^{-3} and are used to catalyse a liquid hydrocarbon reaction. If the density of the liquid hydrocarbon mixture of reactants and products is 800 kgm^{-3} and it has a viscosity of 0.002 Nsm^{-2}, determine the minimum fluidizing velocity of the bed using the Ergun equation.

Solution

A fluidized bed is a vessel or chamber in which solid particles are suspended in an upward flow of a gas (Figure 8.4). The buoyant solid particles therefore behave as though they were in a liquid state. Used extensively in the chemical industry, fluidized beds provide excellent mixing, heat transfer, and mass transfer characteristics. They are used in catalytic reactions where powdered or pelleted catalysts have a high specific surface area. They are also used in furnaces in which coal is combusted in a hot bed of ash or sand through which air is passed. Fluidization permits lower temperatures to be used, thereby avoiding the production of polluting oxides of nitrogen.

The behaviour of a fluidized bed depends on the particle size and the fluidizing gas velocity. When fine particles are fluidized at low gas velocities, the bed expands, but without the formation of bubbles. At higher velocities, at the bubbling regime, there are three distinct zones in the bed: The *grid zone* is located at the bottom of the bed and corresponds to gas penetrating the bed.

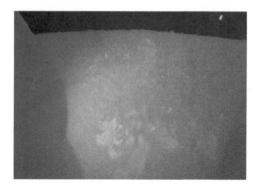

FIGURE 8.4
Cross Section of a Fluidized Bed (Photo from C.J. Schaschke.)

It is dependent on the types of grid used. The *bubbling zone* is where bubbles grow by coalescence and rise to the bed surface where they break. The *freeboard zone* is where some particles are carried above the bed surface and are elutriated from the system while others are returned to the bed (Figure 8.5).

The process of fluidization involves the suspending of solid particles in an upward flow of a fluid. In particulate fluidization involving liquids, each particle behaves individually and collides with others, yet remains a certain distance apart. As the velocity of fluidization is increased, the bed expands. It is

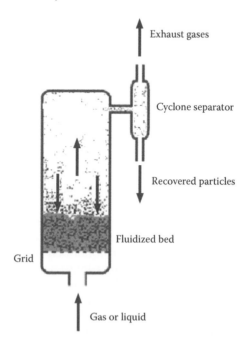

FIGURE 8.5
Bubbling Regimes in a Fluidized Bed

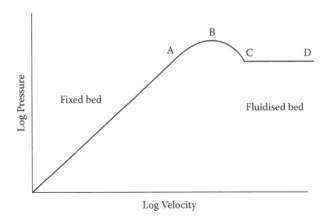

FIGURE 8.6
Pressure Drop across a Bed of Particles

used in the backwashing of filter beds and ion exchange resin beds. In aggregated fluidization involving gases, similar conditions exist up to the point of incipient fluidization. At higher velocities, flow passes through the bed in the form of bubbles, and the bed expands due to the volume of the bubbles.

The behaviour of a fluidized bed depends on the particle size and the gas velocity. When fine particles are fluidized, the bed expands but without the formation of bubbles. This bubbleless regime disappears, however, when the gas velocity is increased above the bubbling velocity.

The support grid has the property of giving a good distribution for fluid flow without generating any appreciable pressure drop. The pressure drop across the support grid can therefore usually be neglected and is usually independent of bed conditions (Figure 8.6).

Zone I: At low flow rates the bed particles remain in position; therefore, bed depth and voidage remain constant since the pressure drop is proportional to the velocity. At point A, the bed starts to expand.

Zone II: At higher fluid velocities, the upward drag of fluid on the particles reduces the particle weight and the particles start to move apart. Interparticulate and wall friction retains the bed in its original position. Under some conditions where particle-to-wall friction is small, the bed may rise as a piston or plug flow. The pressure drop passes through a maximum at B, and then falls slightly to an approximately constant value. This change is due to an increase in the bed voidage and the pressure drop rises more slowly.

Zone III: The particles eventually move sufficiently far apart that they are just touching. Friction disappears, and the pressure drop falls back to a uniform value at the point of incipient fluidization. If the velocity increases further, the pressure drop will remain constant because it is generated by the weight of the bed, which is also constant (line C-D).

Zone IV: Raising the velocity of the gas further may increase the pressure drop due to the friction between the wall of the container but may, however, decrease due to elimination of the particles in the bed, which is time dependent. This condition would be avoided for fluidization because the entire bed would be elutriated from the container. This phenomenon is used in conveying solids in pneumatic conveyors such as grain elevators.

Decreasing the fluid velocity allows the bed to contract where the particles are just touching on each other corresponding to the minimum fluidizing velocity. The voidage has reached the maximum stable value for a fixed bed. Decreasing the fluidizing velocity further allows the bed to be reformed at a pressure across the bed less than that prior to fluidization.

A bed of very small particles will begin to fluidize when the pressure drop across the bed begins to equal the weight of the bed per unit area:

$$\frac{\Delta p}{L} = (1-e)(\rho_P - \rho)g \tag{8.23}$$

The pressure drop over the bed is given by

$$\frac{\Delta p}{L} = \frac{150\mu U(1-e)^2}{(\varphi d_p)^2 e^3} + \frac{1.75\rho U^2(1-e)}{\varphi d_p e^3} \tag{8.24}$$

Known as the Ergun equation after Turkish-born American chemical engineer Sabri Ergun (1918–2006), the incipient point of fluidization corresponds to the highest pressure drop at the minimum fluidization velocity. At the point of fluidization it may be assumed that each spherical particle touches six others in the form of a cube arrangement, such that the voidage is

$$e = 1-\pi/6 = 0.476 \tag{8.25}$$

From Equations 8.23, 8.24, and 8.25,

$$\frac{150(1-0.476)^2 \times 0.002 \times U}{1 \times 0.476^3 \times 0.0015^2} + \frac{1.75(1-0.476) \times 800 \times U^2}{1 \times 0.476^3 \times 0.0015} \tag{8.26}$$

$$= 9.81 \times (1-0.476)(3000-800)$$

Solving gives a fluidization velocity of 0.0157 ms^{-1}.

What Should We Look Out For?

This problem results in the form of a quadratic equation. This is a type of polynomial equation in which the highest power of the unknown variable is two and has the general form

$$ax^2 + bx + c = 0 \tag{8.27}$$

where a, b, and c are constants. In this particular case, one root is positive, and the other is negative and therefore an impossible answer and so is rejected.

What Else Is Interesting?

For particles in the bed with a wide range of particle size distribution, the surface-volume mean d_{sv} should be used. The sphericity or shape factor is best obtained for particles with irregular shapes from the pressure drop through the fixed bed or from the equation for incipient fluidization. Typical values for the sphericity range from 0.7 for coal or angular sand to over 0.9 for round sand or microspherical cracking catalyst. For spherical particles, using the specific area, s, the minimum fluidization velocity is therefore given by the semi-empirical Carman-Kozeny equation:

$$\frac{\dot{Q}}{A} = \frac{1}{Ks^2} \frac{e^3}{(1-e)^2} \frac{\Delta p}{\mu L} \tag{8.28}$$

The equation is typically used to determine the pressure drop through a packed bed of solids and permeability of porous media. Its validity is confined to laminar flow with Reynolds numbers below 1.0, and is named after the Austrian physicist Josef Kozeny (1889–1967), who proposed the equation in 1927; and Philip Carman, who subsequently modified it in 1938 and again in 1956. The constant K varies with bed voidage for different shapes of particles and generally has values in the order of 4.2 to 5.0. For a constant of 5.0, the Carman-Kozeny equation can therefore be usefully approximated to

$$\frac{\Delta p}{L} = \frac{180\mu U (1-e)^2}{d_p^2 e^3} \tag{8.29}$$

The specific surface area of a spherical particle is the ratio of the surface to the volume:

$$s = \frac{4\pi r_p^2}{\frac{4}{3}\pi r_p^3} = \frac{3}{r_p} = \frac{6}{d_p} \tag{8.30}$$

Problem 8.7: Minimum Fluidizing Velocity

A bed of perfectly spherical solid particles with a diameter of 85 μm and density of 3500 kgm^{-3} is fluidized using a gas of density 1.2 kgm^{-3} and a viscosity of 1.9 × 10^{-5} Nsm^{-2}. Determine the minimum fluidizing velocity if a representative sample of particles extracted from the bed occupying a volume of 0.001 m^3 has a mass of 2 kg.

Solution

By assuming laminar flow occurs at onset of fluidization, the pressure drop across the entire bed, the laminar flow part of the Ergun equation is

$$\frac{\Delta p}{L} = (1-e)(\rho_p - \rho)g = \frac{150\mu U(1-e)^2}{d_p^2 e^3} \tag{8.31}$$

The bed voidage, e, is the space between the solid particles expressed as a percentage or fraction of the total volume. The voidage is used to indicate the available space for the flow of a gas or liquid. In terms of volume, the voidage is therefore

$$e = \frac{V_b - \sum V_p}{V_p} = 1 - \frac{\rho}{\rho_p} = 1 - \frac{2000}{3500} = 0.428 \tag{8.32}$$

The void fraction depends on the size and shape of the particles and also their particle size distribution. Light materials are regarded as having bulk densities below 600 kgm^{-3}, while heavy materials have densities in excess of 2000 kgm^{-3}.

Assuming that the gas density is considerably less than that of the particles, the fluidization velocity is

$$U = \frac{\rho_b g d_p^2 e^3}{150\mu(1-e)} = \frac{3500 \times 9.81 \times (8.5 \times 10^{-5})^2 \times 0.428^3}{150 \times 1.9 \times 10^{-5} \times (1-0.428)} = 0.0103 \text{ ms}^{-1} \tag{8.33}$$

What Should We Look Out For?

For small particles only the first (laminar flow) term in the Ergun equation (8.24) is considered in which the Reynolds number has a value below 1. A common difficulty is being able to predict the incipient velocity in which the voidage depends on the particle size, shape, and distribution. An error in estimating the voidage leads to much greater errors in the estimates for the incipient velocity. Available correlations for the voidage are not particularly reliable, and experimental values are therefore usually preferred. An approximate value may be obtained from the bulk density of the solid although the value for the voidage is likely to be greater because the particles tend to pack more closely when crushed upon one another upon filling. An effective way to measure the voidage is to fluidize the bed briefly and allow the bed to resettle.

What Else Is Interesting?

Should weighing or measuring equipment not be available to determine the voidage at the minimum fluidizing conditions, the voidage can alternatively be estimated from a measure of the pressure gradient across the bed. That is, the pressure drop can be measured along with the depth of the expanded bed.

With particulate fluidization, each particle acts individually and has zero effective weight inside the bed since the upward drag of the fluid is equal to the weight of the particle. The particles move around in the bed, colliding with one another and the wall of the container. At higher fluid velocities, the particles move farther apart and the bed expands. That is, the bed depth increases, but the drag force on the particle is still equal to the weight of the particle. Each particle can be considered to be stationary, occupying a cube of bed space (except near the wall of the container and the support plate or grid), and the particle diameter must be small in comparison to the diameter of the container. As a general guide, the minimum particle diameter requirement should be about a hundredth of the diameter of the bed. For streamline flow of the fluid, Stokes' law should apply to each particle in its own "cube" space, although flow is likely to be turbulent for particle Reynolds numbers less than 0.4. Also, the particles collide with each other and the boundary layer around each particle is broken down by the collision but then reforms. There is no viscosity effect at the point of collision; therefore, a large number of collisions have an appreciable effect on the overall viscosity, thereby invalidating Stokes' law.

The main resistance to mass and heat transfer in a fluidized bed is due to the boundary layer. This disappears during the collision of particles such that both mass and heat transfer are greatly increased and chemical reactions in fluidized beds tend to be very rapid. Fluidized beds tend to operate at a constant temperature and uniform concentration within the body of most of the bed.

Aggregative fluidization generally applies to gas fluidized beds in which the conditions are similar up to the point of incipient fluidization, but at higher flows it is found that the bed does not expand uniformly. Instead of

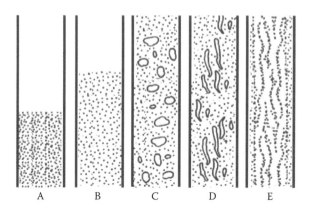

FIGURE 8.7
Fluidized Bed Regimes: A, Fixed Bed; B, Bubbleless; C, Bubbling; D, Turbulent; E, Fast

the particles moving farther apart when the velocity is increased, the bed splits into two parts with one part corresponding to the incipient bed and the other where the gas flow passes through the bed in the form of bubbles which travel faster than the average gas velocity.

Initially, there are a large number of small bubbles, but at a higher gas flow the number of bubbles is reduced, their size increases, and their velocity increases. This is usually referred to as the *bubbling* or *boiling bed* (Figure 8.7).

Up to the point of fluidization, a gas fluidized bed and a liquid fluidized bed are similar and the fluidizing velocity can be calculated from the Carman-Kozeny equation (Equation 8.28). The type of fluidization is determined by the Froude number given by

$$Fr = \frac{U^2}{gd_p} \tag{8.34}$$

where U is the minimum fluid velocity calculated over the entire cross section of the bed, and d_p is the particle diameter (or the diameter corresponding to the surface mean diameter). Since the viscosity of gases is low, the velocity is high and aggregative fluidization occurs for Froude numbers greater than 1 while particulate fluidization occurs for Froude numbers below values of 1.

When the gas velocity increases above the incipient velocity, the additional flow generates bubbles within the bed which rise and burst on the surface of the bed. These bubbles effectively bypass the bed such that any chemical reaction between the bed and the bubble is controlled by diffusion from the bubble into the bed, thereby giving lower mass transfer rates than in the actual bed.

Problem 8.8: Pneumatic Conveyor

Describe the operation of a pneumatic conveyor.

Solution

Pneumatic conveying is the transportation of granular free-flowing particulates such as grains and catalyst particles that are suspended in a fast-moving
flow of air or gas. The particles move freely and can be readily transported to
and from hoppers and silos through pipes and ducts, which may be circular
in cross section or of some other geometry and may be made of various materials including plastic tubing. However, the risk of static electricity, which
may lead to ignition of the particles being transported, needs to be carefully
considered. The flow may be caused by a blower or compressor upstream of
the feed point or may operate with a negative pressure with the blower or
compressor located downstream. For low gauge pressure transportation, a
Roots blower is typically used. This is a type of compressor used to transport
large volumes of gas and consists of two rotating lobe-type rotors which
have a small clearance between the rotors and the casing.

What Should We Look Out For?

There are three identifiable forms of particle transport associated with pneumatic conveying. Particles may generally be kept in suspension where the
solid fraction is kept low, which is known as a dilute or lean phase. High-
velocity transport may result in particle attrition and duct wear. Higher
fractions in horizontal or slightly inclined ducts result in particle settling,
which results in a moving bed phase. Dense phase transportation involves
close-packed slugs of particles. The slugs are propelled along the duct with
a risk of attrition. With low-velocity transport, there is a risk of blockage in
the horizontal ducts.

What Else Is Interesting?

There are a number of identifiable particle flow regimes in pneumatic conveying. Where particles may be of the same size, they may for a given rate of
flow be evenly distributed over the cross section of the duct. Larger particles
within the distribution may, however, tend to flow in the lower section of the
duct or settle on the floor. Slug flow occurs when particles settle on entering the conveyor and settle before accelerating. They create dunes that give
an uneven distribution along the duct. Where the dunes remain stationary,

particles are conveyed above the dunes and are swept from one dune to the next. Particles that settle out near the feed point result in a continuous bed gradually forming along the floor of the duct, which then moves steadily along the duct. The bed of particles may, however, remain stationary, in which the depth of the bed may build up to the point that the duct becomes blocked. Another form of duct blockage occurs with plug flow in which following slug flow, the dunes build up to the point that the duct is entirely blocked.

Problem 8.9: Hydrocyclone Particle Separation

Experimental tests involving spherical particles of a mineral with a density of 2800 kgm^{-3} and a diameter of 5 μm as a suspension in water at a volumetric flow throughput of 200 litres per second show effective separation in a hydrocyclone. Determine the corresponding size cut for a suspension of mineral particles at a volumetric rate of 40 litres per second.

Solution

A hydrocyclone is a type of cyclone device used for the separation of particles suspended in a liquid. It is operated by feeding the liquid suspension tangentially into the top of the device to produce a centrifugal-type vortex action that allows the more dense and coarse particles to be separated in the underflow, while the smaller particles leave through the overflow. Hydrocyclones are typically used in mineral ore processing, drilling mud separation, and oil/water separation in offshore operations. With a low capital cost and inexpensive operational cost, they are effective at separating materials based on small differences in particle size.

Assuming that the particles are sufficiently far apart such that their interaction can be ignored and Stokes' law applies (Equation 8.1), for which the terminal velocity of the mineral particles is

$$U_t = \frac{g(\rho_p - \rho)d_p^2}{18\mu} = \frac{9.81 \times (2800 - 1000) \times (5 \times 10^{-6})^2}{18 \times 0.001} = 2.45 \times 10^{-5} \, \text{ms}^{-1} \quad (8.35)$$

For the mineral-in-oil suspension, the corresponding terminal velocity would have to be

$$U_{t2} = U_t \frac{\dot{Q}_2}{\dot{Q}_1} = 2.45 \times 10^{-5} \times \frac{0.04}{0.2} = 4.9 \times 10^{-6} \, \text{ms}^{-1} \quad (8.36)$$

Again, assuming Stokes' law, the diameter of particles of the second mineral in oil would need to be

$$d_p = \sqrt{\frac{18\mu U_t}{g(\rho_p - \rho)}} = \sqrt{\frac{18 \times 0.01 \times 4.9 \times 10^{-6}}{9.81 \times (1200 - 850)}} = 1.6 \times 10^{-5} \text{ m} \qquad (8.37)$$

What Should We Look Out For?

The experimental test using particles in water may not reflect the performance since the wettability properties of the particles may not be the same. The calculated particle size is based on the assumption of infinite dilution. That is, a particle is not influenced by any surrounding particles. This is very unlikely to be the case in practice, particularly when the hydrocyclone is used to separate particles from a slurry.

What Else Is Interesting?

Hydrocyclones are often used for separating particles suspended in a liquid of lower density by size or density, or generally, by terminal falling velocity. They are also used for the removal of suspended solids from a liquid, such as grit from organic matter, or to remove suspended grit particles from primary sludge in wastewater treatment. They are also effective in separating immiscible liquids of different densities and for the dewatering suspensions to give a more concentrated product, the breaking down of liquid–liquid or gas–liquid dispersions, and the removal of dissolved gases from liquids. Separation is affected in the centrifugal field generated as a result of introducing the feed at a high tangential velocity into the separator. Heavier grit and suspended solids therefore collect on the sides and bottom of the cyclone as a result of centrifugal forces. Scum and lighter solids are removed from the centre through the top of the hydrocyclone.

Problem 8.10: Power Demand in Fluidized Beds

In designing a fluidized bed, there is a requirement for the power demand to be kept to a minimum. The bed is to have a diameter of 1 m and be filled with 1800 kg of spherical particles with a diameter of 2 mm and particle density of 2100 kgm^{-3} to a depth of 1.6 m. Determine the pressure drop over the bed when the rate of flow of fluidizing gas is 0.231 m^{3}s^{-1}, and the power demand if the fluidizing gas has a density of 1.21 kgm^{-3} and a viscosity of 1.42 × 10^{-5} Nsm^{-2}.

Solution

In practice, the volume of the bed of particles is usually fixed. A typical case is a fluidized bed consisting of catalyst particles for which the volume of bed is dependent on the required chemistry of a particular reaction. The power demand for fluidization is therefore the product of the flow rate and pressure drop over the depth of the bed. In practice, the flow rate is likely to have already been fixed according to the mass balance of the process. The bed voidage is found from

$$e = 1 - \frac{\rho}{\rho_p} = 1 - \frac{\frac{m}{AL}}{\rho_b} = 1 - \frac{\frac{1800}{\pi \times 1^2}}{\frac{4}{2100}} = 0.68 \tag{8.38}$$

By combining Equation 8.23 with Equation 8.28, the incipient velocity of the fluidizing gas is

$$U = \frac{\dot{Q}}{A} = \frac{1}{Ks^2} \frac{e^3}{(1-e)} \frac{(\rho_p - \rho)g}{\mu} = \frac{0.231}{\frac{\pi \times 1^2}{4}} = 0.294 \text{ ms}^{-1} \tag{8.39}$$

Assuming that K has a value of 5 and noting that for the spherical particles, the specific surface area of a particle is:

$$s = \frac{6}{d_p} = \frac{6}{0.002} = 3000 \text{ m}^2\text{m}^{-3} \tag{8.40}$$

At the incipient gas velocity, the pressure drop over the bed is given by Equation 8.28, which gives an incipient velocity of

$$U = \frac{1}{5\mu s^2} \frac{e^3}{(1-e)^2} \frac{\Delta p}{L}$$

$$= \frac{1}{5 \times 1.42 \times 10^{-5} \times 3000^2} \times \frac{0.68^3}{(1-0.68)^2} \times \frac{\Delta p}{1.6} = 0.294 \text{ ms}^{-1} \tag{8.41}$$

Solving, the pressure drop is found to be 98 Nm⁻². The power required for the flow of gas through the bed is therefore

$$P_o = \dot{Q}\Delta p = 0.231 \times 98 = 22.6 \text{ W} \tag{8.42}$$

What Should We Look Out For?

By combining the pressure drop over the bed with the incipient velocity equation and power equation, the power required for pumping can alternatively be calculated directly from

$$P_o = \frac{AL}{Ks^2} \frac{e^3 (\rho_p - \rho)^2 g^2}{\mu} \tag{8.43}$$

What Else Is Interesting?

The individual particle characteristics of a fluidized bed include the particle size, shape, density, roughness, and surface area. The bulk properties include voidage or porosity, angle of repose, internal angle of friction heat capacity, and thermal and electrical conductivity. The particle size and size distribution and shape affect the packing and arrangement of the particles in static beds, which influence the pressure drop across the bed and flow of the fluids through the bed. The efficiency of operation of the bed may also be affected. The bulk properties of the particle affect the behaviour of the particulate product during the storage, filling, and discharging of hoppers in the processing and handling of the product. Flow of the bulk solids can also be affected by individual properties of the particulate solid such as the size distribution, shape, and surface roughness. Hygroscopicity is another property of the particulate solids to consider as it can affect the storage of the product.

Problem 8.11: Bubble Nucleation and Growth

A steady stream of bubbles of carbon dioxide gas is found to form on the surface of a glass and to detach and rise in a carbonated beverage of density 1030 kgm^{-3} and viscosity 0.001 Nsm^{-2} for which the following observations and measurements were made:

Time (s)	1.0	2.0	3.0	4.0
Velocity (ms^{-1})	0.0225	0.04	0.0625	0.3

Determine the diameter and age of the bubbles at the point they detach from the nucleation sites on the surface of the glass, and the rate at which carbon dioxide enters the rising bubbles in terms of volume of carbon dioxide (m^3) transferred per unit area (m^2) of bubble if the rate of carbon dioxide transfer is proportional to the surface area of the bubble.

Solution

For liquids that have a high dissolved gas content such as carbon dioxide in water, the formation of bubbles is due to the extent of saturation and sufficient nucleation. Bubbles of gas are formed and grow to a size that is sufficient for them to detach and rise up through the liquid due to buoyancy effects. Their enlargement in size as seen in a stream of nucleating bubbles rising in a champagne glass may initially be thought to be due to the decrease in hydrostatic head. Their enlargement, however, is due largely to the diffusion of carbon dioxide.

If the rate of carbon dioxide gas that transfers into the bubbles changing the volume with time is proportional to the surface area of the bubble, then

$$\frac{dV}{dt} = fa \tag{8.44}$$

The rate of change of bubble radius with time is therefore a constant. That is,

$$\frac{dr}{dt} = f \tag{8.45}$$

From the formation of the bubble, the radius at any time t can be found by integrating Equation 8.45 to give

$$r = f(t + t_o) \tag{8.46}$$

where t_o is the time that the bubble first forms. If the density of the liquid is taken to be considerably greater than that of the gas, then the terminal velocity of the rising bubble of gas rising is approximately given by Stokes' law (see Equation 8.1) expressed using radius rather than diameter as

$$U_t = \frac{2\rho g r^2}{9\mu} \tag{8.47}$$

Therefore,

$$\sqrt{U_t} = f\sqrt{\frac{2\rho g}{9\mu}}(t + t_o) = f\sqrt{\frac{2 \times 1030 \times 9.81}{9 \times 0.001}}(t + t_o) \tag{8.48}$$

Using the data provided gives a straight line (Figure 8.8):

Time (s)	1.0	2.0	3.0	4.0
$\sqrt{\text{Velocity}}$ ($m^{1/2}s^{-1/2}$)	0.15	0.2	0.25	0.3

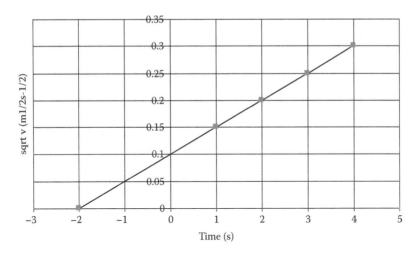

FIGURE 8.8
Bubble Age

The bubbles therefore form 2 seconds before detachment at which time the terminal velocity is 0.01 ms^{-1}. This corresponds to a bubble radius of

$$r = \sqrt{\frac{9\mu U}{2\rho g}} = \sqrt{\frac{9 \times 0.001 \times 0.01}{2 \times 1030 \times 9.81}} = 6.7 \times 10^{-5} \text{ m} \tag{8.49}$$

That is, a bubble diameter of 0.13 mm. The rate of carbon dioxide transfer into the bubble is therefore

$$f = \frac{r}{t + t_o} = \frac{6.7 \times 10^{-5}}{2} = 3.35 \times 10^{-5} \text{m}^3 (CO_2) \text{m}^2 \text{s}^{-1} \tag{8.50}$$

What Should We Look Out For?

The increase in velocity of the bubbles to the surface is therefore neither due to attaining the terminal velocity, which occurs very quickly or to the increase in the bubble diameter due to the reduction in hydrostatic head as they reach the surface. The increase is, in fact, due to the diffusion of carbon dioxide into the bubbles, thereby increasing buoyancy.

What Else Is Interesting?

Bubbles of carbon dioxide form in a carbonated drink due to nucleation that occurs when the pressure is released from a container. Heterogeneous nucleation occurs on a surface. This is best illustrated by adding a salted pea-nut to a carbonated drink. Nucleation also occurs in a boiling liquid when the pressure of the liquid is reduced such that the liquid becomes superheated. Nucleation usually occurs on the heating surface at nucleation sites that are small crevices and surface imperfections. Superheating of a liquid can be achieved even after the liquid has been de-gassed and where the heating sur-faces are clean, smooth, and made of materials that are well wetted by the liquid.

Bubbles can also be formed in liquids as in the case of boiling and compris-ing of bubbles with superheated saturated vapour. As they rise from a heated surface, they can appear to miraculously disappear as they move upward due to the buoyancy effects and into a cooler liquid above.

Problem 8.12: Cyclone Separator

The waste gas from a combustion process contains particles that are sep-arated in a cyclone separator prior to discharge into the atmosphere. The waste gas with a density of 0.9 kgm^{-3} contains a range of particle sizes as shown in Table 8.2. For the particle size distribution and grade efficiency of the cyclone, determine the overall efficiency of the separator.

Solution

A cyclone separator is a device used to separate particles from air or a gas stream and is commonly used in spray drying to collect the dried product. It consists of a vertical cylindrical section with a tapering conical section beneath as shown in Figure 8.9 and is used to separate particle sizes typically ranging from 5 to 200 μm, which enter the separator tangentially with a high velocity of up to 30 ms^{-1}. A vortex flow is therefore created inside the cyclone. The direc-tion of the flow of the outer vortex is downward and close to the cyclone walls, but the flow reverses in the lower part of the cyclone to rise up the centre in a vortex rotating about a turbulent central core, with the gas flow leaving via the outlet in the top. The separated particle-free air or gas leaves the top of the separator. The particles leave via the bottom for collection and recovery.

Taking a basis of 100 g of particles retained, the cumulative mass retained is calculated as shown in Table 8.3, for which the overall efficiency is there-fore 77%.

TABLE 8.2

Particle Distribution Size and Cyclone Grade Efficiency

Size range (μm)	0–5	5–10	10–20	20–40	40–80	80–160
Mass (g) per 100 g	8	14	33	24	12	9
Grade efficiency (%)	25	45	75	90	95	100

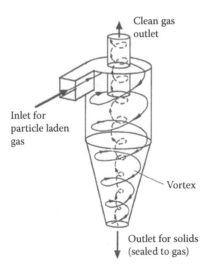

FIGURE 8.9
Cyclone Separator

TABLE 8.3

Particle Size Distribution

Size range (µm)	0–5	5–10	10–20	20–40	40–80	80–160
Retained per 100 g	1.7	5.2	27.2	23.4	10.4	9.0
Cumulative retained per 100 g	1.7	6.9	34.1	57.5	68.0	77.0
Emitted per 100 g	5.25	13.0	19.8	22.1	23.0	23.0

What Should We Look Out For?

The amount emitted below a certain size is based on the cumulative amount, usually expressed as a percentage of the total amount emitted. For example, the amount below 10 µm is therefore 13/23 × 100% = 56.5%.

What Else Is Interesting?

The installation of a cyclone may be an option for a process that produces particles and requires them to be retained to reduce emissions to the environment to meet local authority consent requirements. An example is that the gas density is 0.9 kgm^{-3} and the solids have a concentration of 1 gm^{-3}. This corresponds to 1/0.9 = 1.11 gkg^{-1}, or 1110 ppm by mass. Note that ppm is an abbreviation for parts per million and is often used as a measure of the level of impurities in solids, liquids, and gases. If, for example, a consent level is set at a concentration of 400 ppm, this would correspond to an equivalent concentration of 400/1110 = 0.36 gm^{-3}. If the cyclone emits 23 g per 100 g, the inlet solids concentration would therefore be 0.36/23 × 100% = 1.56 gm^{-3}.

Problem 8.13: Centrifugal Separator

Process water of density 1000 kgm^{-3} and viscosity 0.001 Nsm^{-2} containing spherical particles with a diameter of 30 µm and a density of 2800 kgm^{-3} are to be separated in a continuously operated centrifugal separator. If the bowl of the separator has an inner diameter of 25 cm and rotates at 2000 rpm, for which the inner surface of the centrifuged liquid has a diameter of 5 cm, determine the time it takes to separate the particles.

Solution

A centrifugal separator consists of a fast rotating cylinder or bowl into which a liquid or slurry is fed containing a mixture of solid particles for separation. The liquid forms an annular body against the inner wall of the cylinder or bowl for which the clarified liquid overflows through orifices. The solids settle on the surface of the bowl as a result of centrifugal forces. The solids are pushed along by way of a screw conveyor. The basis of design is the same as the design of settling tanks. In this case, a high centrifugal force on the particles replaces the relatively weaker gravitational force to enhance the speed of separation. Stokes' law (see Equation 8.1) is accordingly modified to

$$U_t = \frac{\omega^2 r\left(\rho_p - \rho\right)d_p^2}{18\mu} \qquad (8.51)$$

The terminal velocity of a particle moving toward the bowl surface is equal to the change in radius with time. The time it takes for a particle initially at the surface liquid to reach the bowl surface can therefore be determined by integrating

$$\int_{r_1}^{r_2} \frac{dr}{r} = \frac{\omega^2\left(\rho_p - \rho\right)d_p^2}{18\mu} \int_0^t dt \qquad (8.52)$$

to give a rearranging of

$$t = \frac{18\mu \ln\left[\dfrac{r_2}{r_1}\right]}{\omega^2\left(\rho_p - \rho\right)d_p^2} = \frac{18\times 10^{-3}\times \ln\left[\dfrac{0.25/2}{0.05/2}\right]}{\left(2\pi\times\dfrac{2000}{60}\right)^2\times(2800-1000)\times\left(30\times 10^{-6}\right)^2} = 0.04 \text{ s} \quad (8.53)$$

For the continuous operation of the centrifuge, the time for a particle to reach the bowl surface is equal to the residence time of the separator.

What Should We Look Out For?

The separation assumes a worst-case scenario in that the particle is required to travel from the inner liquid surface to the wall of the centrifuge. This starting point will not always be the case. It will, however, determine the rate of throughput for the centrifuge. In general, the centrifuge is operated at a fixed speed with a constant input and output feed.

What Else Is Interesting?

The use of centrifugal force on the particles significantly increases the terminal velocity of the particle and hence the rate of separation. This is useful where the particles are very small or where the density difference between the particle and the liquid is very close such that gravity separation would either take too long or the equipment would be infeasibly be too large and costly. Centrifugal separators can also be used to separate suspensions of immiscible liquids such as water dispersed within oils. While being compact in size, they tend to be expensive, and depending on the application, require significant maintenance and downtime.

Further Problems

1. Sketch the flow-pressure drop characteristic of a fluidized bed, and describe the phenomena occurring as the gas flow is increased from zero to well above the fluidization point and then reduced to zero again.

2. Explain the principle of operation of a cyclone separator for use with solid-laden gas streams.

3. Outline with examples the advantages and disadvantages of using fluidized bed reactors to enhance heat and mass transfer in chemical reactions.

4. Together with a sketch, describe the various bubbling zones that exist within a fluidized bed.

5. Determine the power consumption required to pump a liquid with a density of 1100 kgm^{-3} and a viscosity of 1.5 $mNsm^{-2}$ at a rate of 0.250 m^3s^{-1}, through a bed of particles with a diameter of 750×10^{-6} m in a 0.50 m diameter column with a depth of 2 m. The particle bed voidage is 0.45. *Answer*: 842 W

6. Determine the efficiency of a cyclone whose performance to retain particles is as follows:

Size range (μm)	0–5	5–10	10–20	20–40	40–80	80–160
Mass (g) per 100 g	8	16	30	25	13	8
Grade efficiency (%)	25	45	75	90	97	100

Answer: 75%

9

Rheology and Non-Newtonian Fluids

Introduction

The majority of known fluids do not exhibit simple Newtonian behaviour in which the viscosity is independent of shear stress and time, but instead exhibit wide and varied rheological properties. Some flow under the influence of gravity, change shape and form while others remain solid at a particular temperature but are liquid and capable of flow at another, such as waxes. Paints, polymers, and many foods retain their form until a sufficient external force is applied causing them to flow. De-icing fluids sprayed onto the wings of commercial aircraft are formulated to remain in place and prevent ice buildup until the aircraft is at the point of takeoff, where the shear force effects of the air passing over the surface are sufficient to remove them. Toothpaste is designed to remain in place on the toothbrush once squeezed from the tube until sufficient shear is applied by the action of cleaning the teeth.

Fluids that do not exhibit Newtonian behaviour are broadly classified as non-Newtonian fluids. That is, the rate of shear is not directly proportional to the shear stress over all values of shear stress. Instead, the apparent viscosity depends on the shear stress and/or time. Non-Newtonian fluids are further classified as being time dependent, time independent, and viscoelastic (Figure 9.1). While Newtonian fluids are time-independent fluids, non-Newtonian fluids exhibit characteristics where the apparent viscosity either increases or decreases with an increasing shear rate. Examples in which the viscosity decreases with shear stress include polymer melts, paper pulp, wallpaper paste, printing inks, tomato purée, mustard, rubber solutions, and protein concentrations, and are known as pseudoplastic or shear-thinning liquids. Dilatants are fluids in which the apparent viscosity increases with the increasing shear rate. These are more rare and include titanium dioxide suspensions, corn flour/sugar suspensions, cement aggregates, starch solutions, and certain honeys.

The relation between shear stress and the shear rate for time-dependent fluids depends on the history of the time and flow of the fluid. Fluids can be classified as being either thixotropic or antithixotropic (or rheopectic). Rheopexy is a non-Newtonian property of certain fluids that shows a

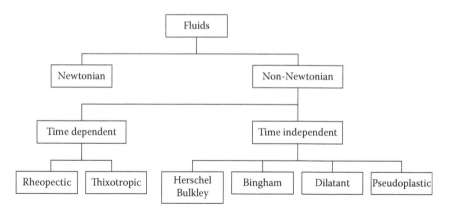

FIGURE 9.1
Classification of Fluids

time-dependent change in viscosity where the fluid thickens more rapidly when stirred or shaken. Rheopectic fluids are comparatively rare and include gypsum suspensions, Bentonite clay suspensions, certain lubricants, and printing inks. Thixotropic fluids exhibit a decrease in shear stress with time for a fixed shear rate. This can be explained by a breakdown in the structure of the fluid as the liquid is sheared. By applying a cyclic variation of applied shear, a hysteresis may be formed in which the fluid properties differ depending on increased or decreased shear. Examples include grease, printing ink, jelly, paint, and drilling mud.

Viscoelastic fluids are non-Newtonian fluids that exhibit the properties of both viscosity and elasticity. Unlike purely viscous fluids where the flow is irreversible, viscoelastic fluids are capable of recovering part of their deformation. Examples include polymeric solutions, partially hydrolysed polymer melts such as polyacrylamide, thick soups, crème fraîche, ice cream, and some melted products such as cheese.

The study of the deformation and flow of fluids is known as rheology and includes the fundamental parameters of elasticity, plasticity, and viscosity. A rheometer is an instrument which is used to obtain the rheological properties of fluids. It involves a small quantity of fluid being entrapped between two geometric surfaces—one surface is static and the other is in motion at a constant speed, which defines the shear rate. The torque to resist the motion defines the shear stress. A limitation with the rotational speed is the need to avoid secondary flow patterns, which give rise to invalid readings.

A parallel-disc, cone-and-plate, and Couette are all forms of rheometers. A viscometer is a particular type of rheometer used to measure viscosity but under no significant applied shear. There are various forms of commonly used viscometers including the Ostwald viscometer, which consists of a bulb into which the liquid is filled and is allowed to flow under gravity through a capillary tube. The time is taken for the meniscus to reach a mark on the capillary from which the viscosity can be determined. Another simple form of a viscometer consists of a spherical ball descending through the liquid under the influence of gravity. The velocity of the falling sphere is based on Stokes' law, and viscosity is determined by timing the sphere at its terminal velocity to descend between two points that are a fixed distance apart.

Problem 9.1: Parallel-Disc Rheometer

A parallel-disc rheometer is used to determine the viscosity of a non-Newtonian fluid. It consists of a flat disc with a diameter of 50 mm that rotates on a flat surface with the liquid being investigated sandwiched between the disc and the surface. For a rotational speed of 600 rpm, which produces a shear stress of 200 Nm⁻², determine the viscosity of a liquid and the torque on the shaft if the clearance between the disc and plate is fixed at 1 mm.

Solution

A parallel-disc rheometer is an instrument used to obtain the rheological properties of non-Newtonian fluids. It consists of a fixed flat surface with another rotating surface held at a fixed elevation above with a sample of the fluid sandwiched between them (Figure 9.2). The rotational speed and distance define the shear rate while the torque to resist the motion defines shear stress. The surface can be heated or cooled to determine the rheological properties as a function of the temperature.

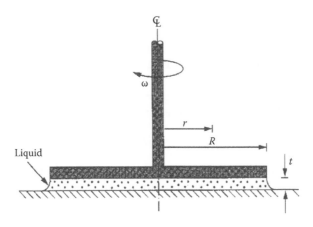

FIGURE 9.2
Parallel-Disc Viscometer

The shear stress is related to the rotational speed as

$$\tau = \mu \frac{2\pi N r}{t} \qquad (9.1)$$

where t is the clearance between the disc and plate. For the non-Newonian fluid being tested, the viscosity is therefore found to be

$$\mu = \frac{\tau t}{2\pi N r} = \frac{200 \times 0.001}{2 \times \pi \times \dfrac{600}{60} \times 0.025} = 0.127 \text{ Nsm}^{-2} \qquad (9.2)$$

The torque on the shaft is

$$T = \frac{\mu \omega \pi}{2t} R^4 = \frac{0.127 \times \left(2 \times \pi \times \dfrac{600}{60}\right) \times \pi}{2 \times 0.001} \times 0.025^4 = 0.0049 \text{ Nm} \qquad (9.3)$$

It is usual to determine the viscosity of a fluid across a range of shear rates and to measure the shear stress. This is carried out by repeating the measurements at different rotational speeds. The apparent viscosity is then determined as the ratio of the shear stress to shear rate, which can then be used to determine the rheological properties of the fluid across a wide range of shear rates. By controlling the temperature of the plate, the rheological properties can also be linked to the influence of the temperature.

What Should We Look Out For?

The calculation is based on assumed laminar flow conditions between the disc and plate. A limitation of the rheometer is that at high rotational speeds, internal vortices known as Taylor vortices can form, which invalidate the measurements. Depending on the design and fluid used, this corresponds to shear rates of around 10,000 s^{-1}.

The shear stress is expressed in terms of the angular velocity of the rotating shaft (rad s^{-1}) and in terms related to the rotational speed (rps):

$$\omega = 2\pi N \qquad (9.4)$$

The torque on the shaft varies across the radius. Combining Equations 9.1 and 9.2, the overall torque is therefore found by integration:

$$T = 2\pi \int_0^R \tau r^2 dr = \frac{2\mu \omega \pi}{t} \int_0^R r^3 dr = \frac{\mu \omega \pi}{2t} R^4 \qquad (9.5)$$

What Else Is Interesting?

In the rheological testing of skin creams, which in practice are applied between the fingers or hand and the skin, a rheometer can simulate the required shearing action. If large shearing effects are required, another type of rheometer called a cone-and-plate rheometer (Problem 9.2) may be used. The rheometer is, however, expected to simulate the shearing of very thin layers without the inclusion of air bubbles and other extraneous materials.

Problem 9.2: Cone-and-Plate Rheometer

A viscous liquid is tested using a cone-and-plate rheometer that has a radius of 60 mm and cone angle of 2.5° for which the following data are obtained:

Speed (rpm)	7.5	15	30	45	60	75	90	105	120
Torque (Nm)	0.022	0.036	0.055	0.067	0.077	0.086	0.093	0.098	0.102

Evaluate the type and properties of the non-Newtonian fluid being tested.

Solution

A cone-and-plate rheometer is an instrument used to measure the rheological properties of fluids (Figure 9.3). It consists of a fixed flat surface with another cone-shaped surface rotating above with a sample of the fluid sandwiched between them. The cone just touches the flat surface. The rotational speed and tapered gap define the shear rate. The torque on the rotating cone

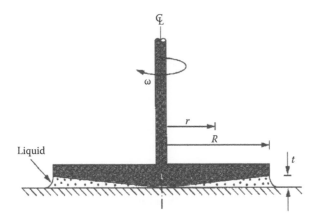

FIGURE 9.3
Cone-and-Plate Rheometer

that resists the motion defines the characteristic shear stress. The surface can be heated or cooled to determine the rheological properties as a function of temperature.

For a cone-and-plate viscometer, the shear rate is given by

$$\gamma = 2\pi N \cot \varphi \tag{9.6}$$

and the shear stress is given by

$$\tau = \frac{3T}{2\pi R^3} \tag{9.7}$$

From the data, the shear stress and shear rate are calculated to be as follows:

Speed (rpm)	7.5	15	30	45	60	75	90	105	120
Shear stress (Nm^{-2})	48.65	79.62	121.6	148.2	170.3	190.2	205.7	216.7	225.5
Shear rate (s^{-1})	18.0	36.0	71.9	107.9	143.8	179.8	215.7	251.7	287.7

The liquid appears to exhibit pseudoplastic characteristics (Figure 9.4) in which the general power law model for the liquid may be assumed:

$$\tau = k\dot{\gamma}^n \tag{9.8}$$

Linearizing:

$$\ln \tau = \ln k + n \ln \dot{\gamma} \tag{9.9}$$

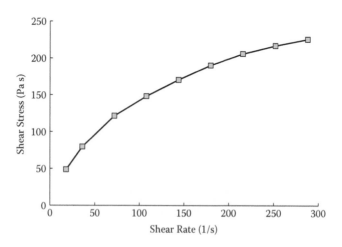

FIGURE 9.4
Rheogram for the Test Liquid

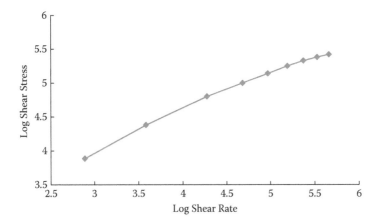

FIGURE 9.5
Linearized Data for the Power Law Model

The linearized data are shown in Figure 9.5, which would appear to reasonbly confirm this model, for which n is found to be 0.55, and k is 2.10 $(Nsm^{-2})^{0.55}$.

What Should We Look Out For?

The cone-and-plate rheometer is designed to operate with a constant shear stress. Operation at high speeds should be avoided as this results in the formation of Taylor vortices leading to inaccurate measurements. The torque varies across the radius of the cone, for which the total torque is found by integration of Equation 9.5.

What Else Is Interesting?

The simplified power law relationship is known as the Ostwald-de Waele equation, which is used to describe non-Newtonian fluids. Depending on the value of the power index, n, the fluid can be classified as being pseudoplastic ($n < 1$), Newtonian ($n = 1$), or dilatant ($n > 1$). It is named after the German chemist Friedrich Wilhelm Ostwald (1853–1932) and British chemist Armand de Waele (1887–1966).

Problem 9.3: Couette Rheometer

Rheological test data for a polymer fluid using a Couette rheometer involves gradually increasing the shear rate to 200 s^{-1} (1) and measuring the shear stress, followed by gradually decreasing the shear rate and measuring the new stress (2). Identify the type and properties of the non-Newtonian fluid.

Shear rate, $\dot{\gamma}$ (s⁻¹)	5	18	45	72	99	113	139	180	200
Shear stress, τ_1 (Nm⁻²)	116	225	440	598	747	817	948	1114	1181
Shear stress, τ_2 (Nm⁻²)	74	203	392	474	684	750	874	1065	1172

Solution

A Couette rheometer is an instrument used to measure the rheological properties of fluids and consists of a cup that fits into a cylindrical bob (Figure 9.6). A sample of the fluid being tested is sandwiched between the bob and the cup. It is operated with either the cup being held and the bob rotated or, more rarely, the bob being held and the cup rotated. The rotational speed and distance between the cup and bob define the shear rate. The torque to resist the motion defines the characteristic shear stress. By heating or cooling the cup, the rheological properties of the fluid can be determined as a function of the temperature.

The apparent viscosity of the fluid is the ratio of the measured shear stress to shear rate:

$$\mu_{app} = \frac{\tau}{\dot{\gamma}} \tag{9.10}$$

The complete range of data is therefore

Shear rate, $\dot{\gamma}$ (s⁻¹)	5	18	45	72	99	113	139	180	200
Viscosity μ_{app1} (Nsm⁻²)	23.2	14.2	8.8	8.3	7.5	7.2	6.8	6.2	5.9
Viscosity μ_{app2} (Nsm⁻²)	14.8	11.3	8.0	7.6	6.9	6.6	6.3	5.9	5.9

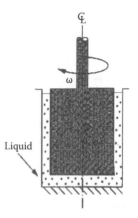

FIGURE 9.6
Couette Rheometer (Photo from C.J. Schaschke.)

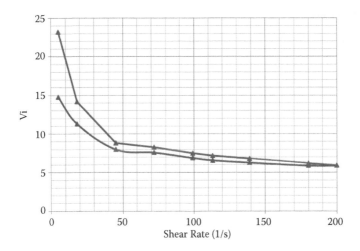

FIGURE 9.7
Variation of Viscosity for the Polymer Fluid

The data are shown in Figure 9.7.

What Should We Look Out For?

In the first measurement with increasing shear rate, the apparent viscosity decreases. This corresponds to a shear-thinning or pseudoplastic fluid. In the second measurement in which the shear rate is successively decreased, the apparent viscosity increases but notably to a lesser extent than in the first set of measurements. This indicates a time dependence or thixotropy. Examples of thixotropic fluids such as certain gels, paints, and lubricants exhibit viscosities that decrease when stress is applied, as in stirring, as well as a time dependency. In contrast, antithixotropic fluids are shear-thickening fluids that thicken with time. The viscosity of such fluids increases when shear stress is applied, as in stirring. Also known as rheopectic fluids, examples include corn flour suspensions, gypsum suspensions, Bentonite clay suspensions, certain lubricants, and printing inks.

What Else Is Interesting?

The Couette rheometer is named after the French physicist Maurice Marie Alfred Couette (1858–1943). Couette flow is a type of flow in which a fluid is sandwiched between two parallel plates—one is stationary and the other is moving at some constant velocity.

Problem 9.4: Power Law Model

Test data from a rheometer of several test fluids are considered to be described by the power law relationship:

$$\tau = k\dot{\gamma}^n \tag{9.11}$$

If the consistency parameter k for all the fluids is 0.01 $(Nsm^{-2})^n$, illustrate the variation of shear stress with shear rate for values of n of 0.9, 1, and 1.1 up to shear rates of 100 s^{-1}.

Solution

The calculation of the shear stress is given in Figure 9.8. With the straight line that corresponds to a Newtonian fluid ($n = 1$), the curve above ($n > 1$) corresponds to a dilatant while the curve below corresponds to a pseudoplastic ($n < 1$).

What Should We Look Out For?

The apparent viscosity is calculated using the determined ratio of shear stress to shear rate (Equation 9.11). This gives the plot as shown in Figure 9.9.

Note that for Newtonian fluids ($n = 1$) there is a straight line with shear rate, while for dilatants ($n > 1$) the curve rises and for pseudoplastics the curve falls. The power law model is valid only within the range of tested data and

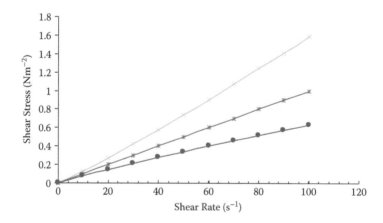

FIGURE 9.8
Power Law Fluids

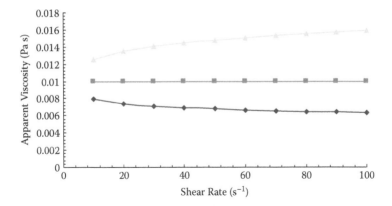

FIGURE 9.9
Variation of Apparent Viscosity for Power Law Fluids

should not be extrapolated beyond the shear rate data used. For example, in the case of predicting data for polymer solutions, errors may occur, where at excessive shear rates, the model ultimately predicts Newtonian behaviour and may wrongly predict an apparent viscosity below that of the solvent.

What Else Is Interesting?

Figure 9.9 is known as a rheogram, which is a graphical plot of rheological parameters which are used in the study of the behaviour of fluids.

Problem 9.5: Rheometer Data Analysis

Test data were obtained from a rheometer for a non-Newtonian liquid. The data consist of the shear stress versus shear rate at a controlled temperature and pressure:

Shear stress (Nm^{-2})	4.5	7.92	11.6	15.5
Shear rate (s^{-1})	2.5	4.0	5.5	7.0

Determine the consistency coefficient k and power law index n for the liquid.

Solution

The data of shear stress versus shear rate present a time-independent power law liquid for which the apparent viscosity decreases with an increasing shear rate. A power law relationship between shear stress and strain rate is

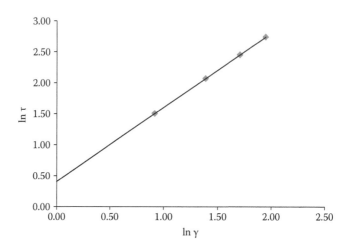

FIGURE 9.10
Linearized Power Law Model

given by Equation 9.11. This can be linearized by using natural logarithms to give

$$\ln\tau = \ln k + n\ln\gamma \tag{9.12}$$

A plot of $\ln\tau$ versus $\ln\gamma$ gives the straight line of gradient n and intercept $\ln k$ (Figure 9.10). The values are therefore calculated to be

$\ln\tau$ (ln Nm^{-2})	1.5	2.07	2.45	2.74
$\ln\Upsilon$ (ln s^{-1})	0.92	1.39	1.70	1.95

From the straight-line relationship, the value of n is 1.2 and k is 1.5 (Nsm^{-2})$^{1.2}$.

What Should We Look Out For?

This type of fluid is therefore a shear-thickening fluid since the viscosity increases with increasing shear rate. That is, n has a value greater than unity.

What Else Is Interesting?

The apparent viscosity of the fluid is the ratio of the shear stress to shear rate. The base SI units for viscosity are kgm^{-1}s^{-2}, although the derived SI units of Nsm^{-2} or Pa s are also acceptable. The poise is a cgs unit, which is also used. It is defined as the tangential force expressed in dynes per square centimetre that is required to maintain the difference in velocity of two parallel plates that sandwich the fluid by a distance of one centimetre at a velocity of one centimetre per second. The centipoise (cP) is more commonly used in which 1 cP is equal to 10^{-3} Pa s.

Problem 9.6: Extrusion of Polymers

A polymer melt at 160°C with a relaxation time of 0.9 s is extruded through a circular die with a diameter of 5 mm. By extrusion through a die with a length of 10 mm, the polymer is found to swell to a diameter of 7 mm, whereas extrusion through a die with a length of 100 mm causes the polymer to swell to a diameter of only 5.2 mm. If the rate of extrusion in both dies is 634.5 mm³s⁻¹, explain why the shortest die has the greatest swell.

Solution

The phenomenon of swelling is known as die swell. This is due to the viscoelastic properties of the polymer. This occurs in a fluid that is partly elastic and partly viscous. The fluid has the ability to remember the deformation to which it has been exposed. In this case, the residence times in the two dies are

$$t_{short} = \frac{V_{short}}{\dot{Q}} = \frac{\frac{\pi d^2}{4} L_{short}}{\dot{Q}} = \frac{\frac{\pi \times 0.005^2}{4} \times 0.01}{634.5} = 0.3 \text{ s}$$

$$\text{(9.13)}$$

$$t_{long} = \frac{V_{long}}{\dot{Q}} = \frac{\frac{\pi d^2}{4} L_{long}}{\dot{Q}} = \frac{\frac{\pi \times 0.005^2}{4} \times 0.1}{634.5} = 3.0 \text{ s}$$

The Deborah number is a dimensionless group used in rheology to classify the behaviour of a fluid which is able to store elastic energy and is defined as the ratio of the fluid characteristic time to the process or observed time where the characteristic time is the stress relaxation time, τ:

$$De = \frac{\tau}{t} \qquad \qquad \text{(9.14)}$$

and that is

$$De_{short} = \frac{\tau}{t} = \frac{0.9}{0.3} = 3$$

$$\text{(9.15)}$$

$$De_{long} = \frac{\tau}{t} = \frac{0.9}{3.0} = 0.3$$

The short die has the highest Deborah number. The elastic energy is therefore greatest in the shortest die in the fluid; hence, there is more swell.

What Should We Look Out For?

The elastic memory of the fluid depends on the relaxation time. The longer residence time in the die, the less the fluid behaves in an elastic manner.

What Else Is Interesting?

Values of the Deborah number approach zero to indicate liquid behaviour, whereas values that approach infinity exhibit solid-like behaviour. The Deborah number was devised by the Israeli scientist Markus Reiner (1886–1976) from the Bible (Judges 5:5) as "The mountains melted from before the Lord." Although a relative number, small values of the Deborah number represent Newtonian flow, while viscous and elastic effects through to elastic and rigid solids correspond to high numbers. Both the numerator and the denominator of Equation 9.15 can effectively alter the Deborah number.

Problem 9.7: Mixing of Non-Newtonian Fluids

A mixing impeller with a diameter of 300 mm with an average shear rate (s^{-1})-to-speed proportionality constant 28 was used to test a Newtonian liquid with a density of 1120 kgm^{-3} and viscosity of 6.12 Nsm^{-2} followed by testing a pseudoplastic fluid. The following results of power uptake against speed were obtained for both fluids:

Newtonian Liquid

Power (W)	5.6	16.7	36.8	88.6	247	986
N (rps)	0.14	0.24	0.35	0.55	0.91	1.73

Pseudoplastic Fluid

Power (W)	18.3	38.4	80.0	127	293
N (rps)	0.29	0.44	0.68	0.89	1.46

If the pseudoplastic liquid can be described by the power law model, determine the parameters for k and n.

Solution

The power required for mixing is related to the reciprocal of the mixing Reynolds number by

$$\frac{P_o}{\rho N^3 D^5} = \frac{K\mu}{\rho N D^2} \qquad (9.16)$$

Therefore,

$$P_o = K\mu N^2 D^3 \qquad (9.17)$$

A plot of power, P_o, with N^2 gives a linear relationship for which K is found to be 1978. Assuming that the same relationship holds with the pseudoplastic, then from Equation 9.16, the apparent viscosity is

$$\mu_{app} = \frac{P_o}{1978 N^2 D^3} \qquad (9.18)$$

The shear rate can be determined from the proportionality constant with rotational speed. Shear stress can therefore be determined using both the apparent viscosity and shear rate (Equation 9.11), and the following data can be obtained:

Apparent viscosity (mNsm^{-2})	4.07	3.71	3.24	3.00	2.57
Shear rate $\dot{\gamma}$ (s^{-1})	8.12	12.32	19.04	24.92	40.88
Shear stress τ (Nm^{-2})	33.05	45.71	61.69	74.76	105.06

For the power law relationship for the pseudoplastic liquid, a plot of the log of the shear rate, $\dot{\gamma}$, with the shear stress, τ, gives a straight line as illustrated in Problem 9.5, for which the k and n are found to be 12.6 and 0.46, respectively.

Problem 9.8: Non-Newtonian Pipe Flow 1

A non-Newtonian fluid flows under isothermal laminar flow conditions through a horizontal pipe. The pipe has an inside diameter of 20 cm and length of 5 m. If the shear stress of the non-Newtonian fluid can be described by the relationship

$$\tau = K \left(\frac{dU_x}{dr} \right)^2$$

where K is constant and has a value of 1000 Ns^{2}m^{-2}, determine the flow rate through the pipe and the average velocity if the measured pressure drop is 10 bar.

Solution

To determine the rate of flow, it is necessary to first determine the variation of the velocity in the cross section of the pipe. A force balance on an element of the fluid along the centreline of a pipe (Figure 9.11) is

$$\Delta p \pi r^2 = \tau 2 \pi r L \qquad (9.19)$$

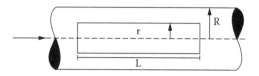

FIGURE 9.11
Pipe Cross Section for Flow

The shear stress is therefore

$$\tau = \frac{r}{2}\frac{\Delta p}{L} \tag{9.20}$$

since the shear stress related to shear rate as

$$\tau = k\left(\frac{dU_x}{dr}\right)^2 \tag{9.21}$$

From Equations 9.20 and 9.21, the change in velocity of the fluid is therefore

$$dU = \left(\frac{\Delta p}{2kL}\right)^{1/2} r^{1/2}dr \tag{9.22}$$

The velocity of the fluid across the cross section can be found by integration:

$$\int_0^{U_x} dU_x = \frac{2}{3}\left(\frac{\Delta p}{2kL}\right)^{1/2}\int_r^R r^{1/2}dr \tag{9.23}$$

to give

$$U_x = \frac{2}{3}\left(\frac{\Delta p}{2kL}\right)^{1/2}\left(R^{3/2} - r^{3/2}\right) \tag{9.24}$$

The volumetric flow rate is found by integrating across the entire cross section:

$$\int_0^Q d\dot{Q} = \int_0^R 2\pi r U_x dr \tag{9.25}$$

to give

$$\dot{Q} = \frac{2\pi}{7}\left(\frac{\Delta p}{2kL}\right)^{1/2} R^{7/2} \tag{9.26}$$

From the data provided, the flow rate is therefore

$$\dot{Q} = \frac{2\pi}{7} \times \left(\frac{10 \times 10^5}{2 \times 1000 \times 5} \right)^{1/2} \times 0.1^{7/2} = 0.002837 \, \text{m}^3\text{s}^{-1} \tag{9.27}$$

The average velocity is taken as the total flow divided by the cross-sectional area of the pipe:

$$\bar{U} = \frac{\dot{Q}}{\pi r^2} = \frac{0.002837}{\pi \times 0.1^2} = 0.09 \, \text{ms}^{-1} \tag{9.28}$$

What Should We Look Out For?

The shear rate is alternatively mathematically expressed as a negative value as

$$\tau = k\left(-\frac{dU_x}{dr} \right) \tag{9.29}$$

The rationale is that the velocity of the fluid decreases from the centre of the pipe to the surface. The negative sign is therefore used to ensure that the shear stress is positive. Equally, the force balance of the fluid pipe may be used mathematically as a negative since the pressure decreases along the length of the pipe.

What Else Is Interesting?

For a general case of non-Newtonian fluids, the shear stress can be given in the general form

$$\tau = k\left(\frac{dU_x}{dr} \right)^n \tag{9.30}$$

The velocity profile, p, of the fluid across the cross section can be found by integration:

$$\int_0^{U_x} dU_x = \left(\frac{\Delta p}{2kL} \right)^{\frac{1}{n}} \int_r^R r^{\frac{1}{n}} dr \tag{9.31}$$

The velocity profile is found by using the boundary conditions of no flow at the wall. Integration therefore gives

$$U_x = \frac{n}{n+1}\left(\frac{\Delta p}{2kL} \right)^{1/2}\left(R^{\frac{n+1}{n}} - r^{\frac{n+1}{n}} \right) \tag{9.32}$$

Integration over the cross section provides the equation for volumetric flow rate as

$$\dot{Q} = \int_0^Q 2\pi r U_x dr \tag{9.33}$$

giving

$$\dot{Q} = \frac{\pi n}{3n+1} \left(\frac{\Delta p}{2kL} \right)^{\frac{1}{n}} R^{\frac{3n+1}{n}} \tag{9.34}$$

The average velocity can therefore be given as

$$\bar{U} = \frac{\dot{Q}}{\pi R^2} = \frac{n}{3n+1} \left(\frac{\Delta p}{2kL} \right)^{\frac{1}{n}} R^{\frac{n+1}{n}} \tag{9.35}$$

Problem 9.9: Non-Newtonian Fluid Flow

A non-Newtonian fluid that can be described by a power law model for which n is 0.2 and k is 5 (kPa s)^n is to be pumped through a pipe of length 10 m and inside diameter of 20 mm. Determine the radius required for another pipe of 20 m in length to carry the fluid at the same rate with the same pressure drop.

Solution

The rate of flow for the non-Newtonian fluid is given by the generalised Equation 9.34:

$$\dot{Q} = \frac{\pi n}{3n+1} \left(\frac{\Delta p}{2kL} \right)^{\frac{1}{n}} R^{\frac{3n+1}{n}} \tag{9.36}$$

For the two pipes with the same pressure drop and flow rate, it follows that

$$\frac{R_1^{\frac{3n+1}{n}}}{L_1^{\frac{1}{n}}} = \frac{R_2^{\frac{3n+1}{n}}}{L_2^{\frac{1}{n}}} \tag{9.37}$$

The new pipe radius is therefore

$$R_2 = R_1 \left(\frac{L_2}{L_1} \right)^{\frac{1}{3n+1}} = 0.01 \times \left(\frac{20}{10} \right)^{\frac{1}{1.6}} = 0.015\,\text{m} \tag{9.38}$$

that is, a required inside pipe diameter of 30 mm.

What Should We Look Out For?

From Equation 9.35, the pressure drop along the pipe for a power law fluid can be given by

$$\frac{\Delta p}{L} = \frac{2k\bar{U}^2}{d^{n+1}} \left(\frac{6n+1}{n} \right)^n \tag{9.39}$$

for which the general equation for the friction factor can be given by

$$f = \frac{\Delta p d}{2\rho L \bar{U}^2} \tag{9.40}$$

The friction factor for a power law non-Newtonian fluid can therefore be conveniently expressed as

$$f = \frac{2k\bar{U}^{-2-n}}{\rho d^n} \left(\frac{6n+1}{n} \right)^n \tag{9.41}$$

What Else Is Interesting?

For the case of laminar flow in which the friction factor is related to the Reynolds number by

$$f = \frac{16}{\text{Re}} \tag{9.42}$$

then

$$\text{Re} = \frac{\rho \bar{U}^{-2-n} d^n}{8^{n-1} k} \left(\frac{4n}{3n+1} \right)^n \tag{9.43}$$

In the special case where n has a value of 1 and $k = \mu$, the equation reduces to that of a Newtonian fluid.

Problem 9.10: Non-Newtonian Pipe Flow 2

A gelatinous foodstuff is to be conveyed from a holding vessel to a packaging line along a 10-m straight pipe run using pressurised sterile air above the material as a propellant to provide a volumetric flow rate to the packaging line of 5 m³h⁻¹. Care is required in the transportation since the texture of the foodstuff is known to be permanently degraded if the bulk of the material is subjected to high shear rates that exceed 250 s⁻¹. Using data that were obtained with a laboratory test using a capillary viscometer with a diameter of 2.5 mm and length of 500 mm, determine the minimum acceptable inside diameter of the transfer pipe and the pressure drop along the pipe if this diameter is to be used.

$\dot{Q}$ (m³ × 10⁻⁶) min⁻¹	1.70	3.70	9.78	15.45	19.80
Δp (Nm⁻² × 10⁻⁵)	1.30	2.17	4.40	5.78	6.88

Solution

Assuming that the gelatinous foodstuff is non-Newtonian, the shear rate is given by

$$\dot{\gamma} = \frac{\dot{Q}}{\pi R^3}\left(3 + \frac{d\ln\dot{Q}}{d\ln\Delta p}\right) \tag{9.44}$$

A plot of $\ln\dot{Q}$ against $\ln\Delta p$ gives a straight line with a gradient of 1.47. The minimum pipe radius is found to be

$$R = \sqrt[3]{\frac{\dot{Q}}{\pi\dot{\gamma}}\left(3 + \frac{d\ln\dot{Q}}{d\ln\Delta p}\right)} = \sqrt[3]{\frac{5/3600}{\pi\times 250}(3+1.47)} = 0.02\ \text{m} \tag{9.45}$$

From a force balance on the foodstuff flowing in the pipe, the wall shear stress is

$$\tau_w = \frac{\Delta p R}{2L} \tag{9.46}$$

This value is the same for both the pipe and the capillary viscometer. Using the data for the viscometer, the pressure drop along the capillary would therefore correspond to 1810 × 10⁵ Nm⁻², so the pressure drop along the pipe would be

$$\Delta p_{pipe} = \frac{L_{pipe}R_{cap}}{L_{cap}R_{pipe}}\Delta p_{cap} = \frac{10\times 2.5\times 10^{-3}}{0.5\times 0.02}\times 1810\times 10^5 = 452.5\times 10^5\,\text{Nm}^{-2} \tag{9.47}$$

What Should We Look Out For?

In practice, a wider-bore pipe would be used to ensure that the food is not degraded and the pressure drop is reduced.

What Else Is Interesting?

Capillary viscometers are based on the measurement of time for the fluid to flow under the influence of gravity through a narrow tube. While the gravitation force is constant and gives reliable results, the force is too small to test highly viscous fluids and non-Newtonian fluids or fluids which have a yield shear stress. Capillaries of differing diameters are therefore required to cover a range of viscosities. A range of types is used including direct-flow and reverse-flow capillaries. In direct-flow capillaries, the sample reservoir that contains the fluid is located below the measuring marks, while in reverse-flow capillary viscometers the reservoir sits above the marks. Reverse-flow capillaries are used for testing opaque liquids and can have a third measuring mark. This allows for the measurement of two flow times and can be used to define the measurement's determinability.

Further Problems

1. From a molecular point of view, provide an explanation the change of viscosity with increasing shear rates for pseudoplastic and dilatant fluids.

2. Explain what is generally meant by a non-Newtonian fluid.

3. A parallel-plate device is used to determine the viscosity of a non-Newtonian fluid. The gap is set at 0.7 mm, and the radius is 25 mm. If the measured torque is 0.004 Nm for which the rotational speed is 10 rps determine the viscosity. *Answer*: 0.073 Nsm^{-2}

4. A non-Newtonian fluid is known to exhibit either a power law or Bingham plastic behaviour over a shear rate range of 10 s^{-1} and 100 s^{-1}. If the yield stress is 20 Nm^{-2} and the plastic viscosity is 0.12 Nsm^{-2}, determine the parameters k and n in the Ostwald-de Waele equation. *Answer*: 14 and 0.179

5. A toothpaste is to be dispensed from a tube that has a cylindrical nozzle with an inside diameter of 6 mm. If the yield stress of the toothpaste is 3 kNm^{-2}, determine the pressure that is required to be applied in the tube in order for the toothpaste to flow from the nozzle. *Answer*: 20 kNm^{-2}

6. A non-Newtonian fluid is transported through a pipe length of 5 m and inside diameter of 200 mm. If the pressure drop across the pipe is 88 Nm^{-2}, determine the volumetric flow rate and average velocity if the fluid can be described by the power law model for which the consistency parameter is 2000 Ns^{3}m^{-2} and n is 3. *Answer:* 0.0033 m^{3}s^{-1}, 0.1 ms^{-1}

7. A Bingham fluid is tested using a rheometer and gives the following experimental data:

Shear rate (s^{-1})	0.21	0.95	1.38	1.92	2.63	3.05	3.67	4.32	5.07
Shear stress (Nm^{-2})	69.9	75.4	78.8	82.5	88.4	91.2	96.2	101.2	106.9

If the fluid can be described by $\tau = \tau_o + \mu\dot{\gamma}$, determine the yield stress and viscosity.

Answer: 68.1 Nm^{-2}, 7.63 Nsm^{-2}

Further Reading

Acheson, D.J. 1996. *Elementary Fluid Dynamics*, 4th ed. Oxford University Press, Oxford.

Al-Sheikh, J.N., Saunders, D.E. and Brodkey, R.S. 1970. Prediction of flow patterns in horizontal two-phase type flow. *Can. J. Chem. Eng.* 48:21–29.

Andeen, G.B. and Griffith, P. 1968. Momentum flux in two-phase flow, *J. Heat Transfer* 90(2):211–222.

Baar, R. and Riess, W. 1997. Two-phase flow velocimetry measurements by conductive-correlative method. *Flow Meas. Instrum.* 8(1):1–6.

Baker, O. 1954a. Designing for simultaneous flow for oil and gas. *Oil Gas J.* 12:185.

Baker, O. 1954b. Simultaneous flow of oil and gas. *Oil Gas J.* 53(12):185–195.

Bankoff, S.G. 1960. A variable-density single-fluid model for two-phase flow with particular reference to steam-water flow. *Trans. ASME, Series C, J. Heat Transfer* 82:265–272.

Barnea, D. and Taitel, Y. 1985. Flow-pattern transitions in two-phase gas-liquid flows, Chapter 16 in *Encyclopaedia of Fluid Mechanics*, vol. 3 (edited by Cheremisinoff, N.). Gulf Publishing, Houston, TX.

Baroczy, C.J. 1966. A systematic correlation for two-phase pressure drop. *Chem. Eng. Prog. Symp. Ser.* 62(64):232–249.

Beggs, H.D. and Brill, J.P. 1973. A study of two-phase flow in inclines pipes. *J. Petroleum Tech.*, 25:607–617.

Bhaga, D. and Weber, M.E. 1972. Holdup in vertical two-and three-phase flow. *Can. J. Chem. Eng.* 50:323.

Boxer, G. 1988. *Fluid Mechanics*. Macmillan, London.

Butterworth, D. 1975. A comparison of some void fraction relationships for gas–liquid flow. *Int. Multiphase Flow* 1:845–850.

Butterworth, D. and Hewitt, G.F. 1977. *Two-Phase Flow and Heat Transfer*. Oxford University Press, Oxford, UK.

Chen, J.J.J. and Spedding, P.L. 1983. An analysis of holdup in horizontal two-phase gas liquid flow. *Int. J. Multiphase Flow* 9(2):147–159.

Chen, J.J.J. and Spedding, P.L. 1984. Holdup in horizontal gas-liquid flow, in *Multiphase Flow and Heat Transfer*, vol III, part A: Fundamentals (edited by Veziroglu, T.N. and Bergles, A.E.). Elsevier, Amsterdam.

Cheremisinoff, N.P. and Gupta, R. 1983. *Handbook of Fluids in Motion*. Ann Arbor Science, Michigan.

Chisholm, D. 1967. A Theoretical basis for the Lockhart-Martinelli correlation for two-phase flow. *Int. J. Heat Mass Transfer* 10:1767–1778.

Churchill, S.W. 1977. Friction-factor equation spans all fluid-flow regimes. *Chem. Eng.* 84:91–92.

Clark, N.N. and Flemmer, R.L. 1985. Predicting the holdup in two-phase bubble upflow and downflow using the Zuber and Findley drift-flux model, *AIChEJ* 31(3):500–503.

Currie, I.G. 1993. *Fundamentals of Fluid Mechanics*, 2nd ed. McGraw-Hill, New York.

Darby, R. 1996. *Chemical Engineering Fluid Dynamics*. Marcel Dekker, New York.

Douglas, J.F., Gasiorek, J.M. and Swaffield. 1996. *Fluid Mechanics*, 3rd ed. Longman, Harlow, UK.

Douglas, J.F. and Mathews, R.D. 1996a. *Solving Problems in Fluid Mechanics*, vol. 1, 3rd ed. Longman, Harlow, UK.

Douglas, J.F. and Mathews, R.D. 1996b. *Solving Problems in Fluid Mechanics*, vol. 2, 3rd ed. Longman, Harlow, UK.

Fisher, S.A. and Yu, S.K.W. 1975. Dryout in serpentine evaporators. *Int. J. Multiphase Flow* 1:771–791.

Fox, R.W. and McDonald, A.I. 1994. *Introduction to Fluid Mechanics*, 4th ed. John Wiley, New York.

French, R.H. 1994. *Open Channel Hydraulics*. McGraw-Hill, New York.

Golan, L.P. and Stenning, A.H. 1969. Two-phase vertical flow maps. *Proc. Inst. Mech. Engrs.* 184(3C):108–114.

Grace, J.R., Wairegi, T. and Nguyan, T.H. 1976. Shapes and velocities of single drops and bubbles moving freely through immiscible liquids. *Trans. Inst. Chem. Eng.* 54:167–173.

Granet, I. 1996. *Fluid Mechanics*, 4th ed. Prentice Hall, Englewood Cliffs, NJ.

Griffith, P. and Wallis, G.B. 1961. Two-phase slug flow. *Trans* ASME (Series C, *J. Heat Transfer*) 83:307–318.

Hetsroni, G. 1982. *Handbook of Multi-Phase Systems*. McGraw-Hill, New York.

Hewitt, G.F. 1982a. Pressure drop, Chapter 2.2 in *Handbook of Multiphase Systems* (edited by Hetsroni, G.). McGraw-Hill, New York.

Hewitt, G.F. 1982b. Void fraction, Chapter 2.3 in *Handbook of Multiphase Systems* (edited by Hetsroni, G.). McGraw-Hill, New York.

Hewitt, G.F. 1984. Two-phase flow through orifices, valves, bends and other singularities. *Proceedings of the Eighth Lecture Series on Two-Phase Flow*, University of Trondheim, pp. 163–198.

Hughes, W.F. and Brighton, J.A. 1991. *Schuam's Theory and Problems—Fluid Dynamics*, 2nd ed. McGraw-Hill, New York.

Hughmark, G.A. 1982. Holdup in gas-liquid flow. *Chem. Eng. Prog.* 58:4.

Levy, S. 1980. Steam-slip—Theoretical prediction from a momentum model. *Trans. ASME J. Heat Transfer* 82:113–124.

Ligget, J.A. 1994. *Fluid Mechanics*. McGraw-Hill, New York.

Lockhart, R.W. and Martinelli, R.C. 1949. Proposed correlation of data for isothermal, two-phase, two-component flow in pipes. *Chem. Eng. Prog.* 45:39–48.

Manahane, J.M., Gregory, G.A. and Aziz, K.A. 1974. A flow pattern map for gas liquid flow in horizontal pipes. *Int. J. Multiphase Flow* 1:537–553.

Martinelli, R.C. and Nelson, D.B. 1948. Prediction of pressure drop during forced circulation boiling of water. *Trans. ASME* 70:695.

Massey, B.S. 1997. *Mechanics of Fluids*, 6th ed. Chapman & Hall, London.

Mishima, K. and Hibiki, T. 1996. Some characteristics of air–water two-phase flow in small diameter vertical tubes. *Int. J. Multiphase Flow* 22:703–712.

Moody, L.F. 1944. Friction factors in pipe flow. *Trans. ASME* 66:671.

Mott, R.L. 1994. *Applied Fluid Mechanics*, 4th ed. Prentice Hall, Englewood Cliffs, NJ.

Munson, B.R., Young, D.F. and Okiishi, T.H. 1994. *Fundamentals of Fluid Mechanics*, 2nd ed. John Wiley, New York.

Muzychka, Y.S. and Awad, M.M. 2010. Asymptotic feneralizations of the Lockhart-Martinelli method for two-phase flows. *J. Fluids Eng.* 132.

Nikuradse, J. 1933. *Forschungsheft*, 301.

Oshinowo, T. and Charles, M.E. 1974. Vertical two-phase flow, Part 1: Flow pattern correlations. *Can. J. Chem. Eng.* 52:25–35.

Penney, W.R. 1978. Inert gas in liquid mars pump performance. *Chem. Eng.*, July 3:63.

Premoli, A., Francesco, D. and Prina, A. 1971. A dimensionless correlation for determining the density of two-phase mixtures. *Termotecnica* 25:17–26.

Prueli, D. and Gutfinger, G. 1997. *Fluid Mechanics.* Cambridge University Press, Cambridge, UK.

Reader-Harris, M.J., Brunton, W.C., Gibson, J.J., Hodges, D. and Nicholson, I.G. 2001. Discharge coefficients of Venturi tubes with standard and non-standard convergent angles. *Flow Meas. Instrum.* 12:135–145.

Roberson, J.A. and Crowe, C.T. 1997. *Engineering Fluid Mechanics*, 6th ed. John Wiley, New York.

Sakaguchi, T., Akagaura, K., Hamaguchi, H., Imoto, M. and Ishida, S. 1979. Flow regime maps for developing steady air-water two-phase flow in horizontal tubes. *Mem. Rac. Eng. Kobe. Univ.* 25:191–202.

Schaschke, C.J. 1998. *Fluid Mechanics: Worked Examples for Engineers.* IChemE, Rugby, UK.

Sharpe, G.T. 1994. *Solving Problems in Fluid Dynamics.* Longman, Harlow, UK.

Smith, S.L. 1971. Void fraction in two-phase flow—A correlation based upon an equal velocity head model. *Heat Fluid Flow* 1(1):22–39.

Spedding, P.L. and Chen, J.J.J. 1984. Hold-up in two-phase flow. *Int. J. Multiphase Flow* 10(3):307–339.

Spedding, P.L. et al. 1982. Pressure drop in two-phase, gas-liquid flow in inclined pipes. *Int. J. Multiphase Flow* 8(4):407–421.

Street, R.L., Watters, G.Z. and Vennard, J.K. 1996. *Elementary Fluid Mechanics*, 7th ed. John Wiley, New York.

Streeter, V.L. and Wylie, E.B. 1983. *Fluid Mechanics.* McGraw-Hill, New York.

Taitel, Y. and Dukler, A.E. 1976. A model for predicting flow regime transitions in horizontal and near horizontal gas-liquid flow. *AIChEJ* 22:47–55.

Thom, J.R.S. 1964. Prediction of pressure drop during forced circulation boiling of water. *Int. J. Heat Mass Transfer* 7:709–724.

Tritton, D.J. 1997. *Physical Fluid Dynamics*, 2nd ed. Oxford Series Publication, Oxford, UK.

Tsai, M.J. 1982. Accounting for dissolved gases in pump design. *Chem. Eng.* July 26:65.

Turner, I.C. 1996. *Engineering Applications of Pneumatics and Hydraulics.* Arnold, London.

Vardy, A. 1990. *Fluid Principles.* McGraw-Hill, New York.

Wallis, G.B. 1969. *One-Dimensional Two-Phase Flow.* McGraw-Hill, New York.

Weisman, J. and Kang, S.Y. 1981. Flow-pattern transitions in vertical and upwardly inclined lines. *Int. J. Multiphase Flow* 7:271.

West, J.B. 2013. Torricelli and the ocean of air: The first measurement of barometric pressure. *Physiology*, March; 28(2):66–73.

White, F.M. 1994. *Fluid Mechanics*, 3rd ed. McGraw-Hill, New York.

Widden, M. 1996. *Fluid Mechanics.* Macmillan, London.

Yamazaki, Y. and Yamaguchi, K. 1979. Characteristics of cocurrent two-phase downflow in tubes. *J. Nucl. Sci. Technol.* 16:245–255.

Young, D.F., Munson, B.R. and Okiishi, T.H. 1997. *A Brief Introduction to Fluid Mechanics.* John Wiley, New York.

Zivi, S.M. 1964. Estimation of steady-state steam void fraction by means of the principle of minimum entropy function. *Trans ASME, J. Heat Transfer* 86:247–252.

Index